TRIUMPH

Books by Jeremy Schaap

CINDERELLA MAN:
JAMES J. BRADDOCK, MAX BAER,
AND THE GREATEST
UPSET IN BOXING HISTORY

TRIUMPH:
THE UNTOLD STORY OF JESSE OWENS
AND HITLER'S OLYMPICS

TRIUMPH

THE UNTOLD STORY OF JESSE OWENS AND HITLER'S OLYMPICS

Jeremy Schaap

A MARINER BOOK
HOUGHTON MIFFLIN COMPANY
Boston   New York

First Mariner Books edition 2008

Copyright © 2007 by Jeremy Schaap

For information about permission to reproduce
selections from this book, write to Permissions,
Houghton Mifflin Company, 215 Park Avenue South,
New York, NY 10003.

www.houghtonmifflinbooks.com

Library of Congress Cataloging-in-Publication Data

Schaap, Jeremy.
Triumph : the untold story of Jesse Owens and Hitler's
Olympics / Jeremy Schaap.
p. cm.
Includes bibliographical references and index.
ISBN-13: 978-0-618-68822-7
ISBN-10: 0-618-68822-6
1. Owens, Jesse, 1913–1980   2. Track and field athletes — United
States — Biography. 3. African American athletes — Biogra-
phy. 4. Jewish athletes — United States — Biography. 5. Glick-
man, Marty, 1917–2001. 6. Stoller, Sam 1915–1983. 7. Olympics
— Participation, American. 8. Olympic Games (11th : 1936 :
Berlin, Germany) 9. National socialism — Philosophy.
10. Racism — Germany — History — 20th century. I. Title.
GV697.O9S33 2007
796.42092 — dc22   [B]   2006026926
ISBN-13: 978-0-618-91910-9 (pbk.)
ISBN-10: 0-618-91910-4 (pbk.)

Book design by Melissa Lotfy

PRINTED IN THE UNITED STATES OF AMERICA

MP 10 9 8 7 6 5 4 3 2 1

Text from *Leni Riefenstahl*, a memoir by Leni Riefenstahl, copyright © 1993
by the author, and reprinted by permission of St. Martin's Press, LLC.

For Lester Gottlieb,
a consummate grandfather
and gentleman

He had great power in his legs . . . he had blinding speed . . .
and his style was flawless — with no sign of extra effort.
Jesse was as smooth as the west wind.

—GRANTLAND RICE

Contents

x • Contents

Prologue

JUST BEFORE 9:30 P.M. central time on September 23, 1955, in a handsome townhouse on Chicago's South Side, James Cleveland Owens slipped into a tweed jacket and sat down in a straight-backed chair. As he smoothed out his pencil mustache and slicked back his hair — what little was left of it — a dozen technicians put the finishing touches on what had been an all-day job, wiring and lighting the Owens home. In a few minutes, Owens would be talking live on national television with Edward R. Murrow of CBS, on his celebrity interview show *Person to Person*. More than 20 million Americans would watch as Murrow spoke from a studio in New York via satellite, first with Owens and his family, and then, in the second half of the show, with Leonard Bernstein and his.

A forty-two-year-old father of three, Jesse Owens weighed twenty-five pounds more than he had in Berlin in 1936, when he had turned in the most indelible performance ever at the Olympic games. In his conservative jacket, flannel slacks, white shirt, and dark tie, he could have passed for a fifty-year-old. Not that he wasn't in superb shape. He was. In fact, just a few months earlier he had run 100 yards in 9.9 seconds, less than a second slower than his personal best. He still held the world record in both the broad jump (now called the long jump) and the 4 x 100-meter relay — though both records had been set in the mid-1930s.

For his part, Murrow was readying himself for another half-hour of banalities. No one confused *Person to Person* with *See It Now,* Murrow's other show on CBS, the one on which eighteen months earlier he had neutered Senator Joseph McCarthy. Despite the fluff, Murrow was eager to speak with Owens, whose legend had grown significantly since 1936. Here, Murrow thought, was a legitimate American hero, the man who had humbled the Third Reich.

For Owens, the appearance with Murrow was emblematic of his enhanced stature. In the first fifteen years after his athletic career ended, he had struggled to find his way, professionally and financially. He made more money than the vast majority of his fellow Americans — in the dry-cleaning business, at Ford Motors, working for the state of Illinois — but the windfall he expected in the aftermath of his Olympic heroics never materialized. Banned from amateur competition after an imbroglio with American track officials, he had raced against horses — most famously in Havana, in December 1936, defeating Julio McCaw, a five-year-old bay gelding, after the horse spotted him a 40-yard advantage. In 1938, on the occasion of the first night baseball game at Ebbets Field, he raced two speedy major-league outfielders, spotting them several yards. He barnstormed with a black baseball team and campaigned for the Republican presidential candidate Alf Landon. In countless ways, he sold himself — but he never had much to show for it. Until now.

By the time Owens sat down to speak with Murrow, he was well on his way to becoming an institution — the Jesse Owens who would spend the rest of his life telling his story to appreciative audiences around the world, the Jesse Owens who could have been a hero from Horatio Alger, if Alger's heroes had not all been white. In the years after his Olympic victories, his achievements in Berlin had been overshadowed by World War II. But by 1955, at the end of the first decade of the cold war, he was finally getting his due. He was in demand as a banquet speaker and making good money because he had become useful — to industry and government — as a symbol of the opportunities America promised and sometimes delivered. To the delight of white America and most of black America, he disputed the sentiments of Paul Robeson, the All-American football player turned actor/singer, who famously suggested that African-Americans would not and should not fight for the United States in the event of war with the Soviet Union. Owens, in contrast, held himself out as an example of what black Americans could achieve, despite the indignities and slights he had suffered his entire life. He agreed with Jackie Robinson, who in his 1949 testimony before the House Un-American Activities Committee had said that blacks had too much invested in the American experiment to support its enemies.

Just a few days after his appearance on *Person to Person*, Owens was to embark, at the behest of the State Department, on a goodwill tour of Singapore, Malaya, the Philippines, and India. As A. M. Rosenthal, then the *New York Times* correspondent in South Asia, put it, Owens's mission was "to make friends for the United States." Having fought the fascists with his fleetness of foot, he would now fight the Communists with his charm and rhetoric — even though some Indian writers, unversed in the annals of the Olympics, confused him with Sir Owen Dixon, an Australian judge and United Nations mediator.

Before the long ride to the subcontinent, though, there was the interview with Murrow, whose fondness for bespoke tailoring matched his own. Finally, at 9:30, with a cigarette clenched in his left hand, Murrow began the interview.

"Jesse Owens," he said, "is generally recognized as the greatest track star of the last half-century. His performance in Berlin stands unmatched in modern times. Statistics will never indicate Adolf Hitler's reaction as he watched a twenty-three-year-old boy from Danville, Alabama, run the athletes of the master race right into the ground." Owens, whose politeness was among his defining characteristics, declined to correct Murrow by pointing out that he had been twenty-two, not twenty-three, and was from Oakville, Alabama, not Danville. He simply smiled and waited for the questions he knew were coming, the questions that always came.

After several minutes of amiable chatter — "You look to be in almost good enough condition to get out your old track shoes again" — and the introduction of Owens's wife and three pretty daughters, Murrow offered him the opportunity to talk about the games of the Eleventh Olympiad. "Jesse Owens," he said, "what's your warmest memory of that August of 1936?"

Owens had been asked this question, or its variants, perhaps hundreds of times. He did not hesitate. "I remember a boy," he said, his accent betraying no hint of his southern roots, "that I competed against in the broad jump — a boy with whom I built a friendship — and we corresponded for a number of years, and then the war broke out, and I didn't hear any more from him at all."

Owens looked down and away from the camera. The boy he was referring to was Luz Long, the silver medalist, a pure-blooded Aryan from Leipzig who had helped him reach the broad-jump finals when he had been on the verge of disqualifying. Composing himself, Owens talked about Long's son Kai — Owens and Kai had met in 1951 — and then about winning the 100-meter dash. But he had not yet answered Murrow's question.

"I think that the greatest moment that a person can have is to stand on a victory stand," he said, "far away from home, and then, from the distance you can hear the strains of 'The Star-Spangled Banner,' and then suddenly you make a left turn and you see the Stars and Stripes rising higher and higher, and the higher the Stars and Stripes rose the louder the strains of the Star-Spangled Banner would be heard. I think that's the greatest moment of my whole athletic career."

Finished, he smiled, looking slightly off-camera.

"Thank you very much, Jesse Owens," Murrow said, taking a deep drag. "In just a moment, we'll take you for a visit with Leonard Bernstein and his wife, Felicia Montealegre."

Now Owens rose from his chair and dug into his pants pockets. Unlike Murrow, he had not dared to smoke on camera. Acutely conscious of his image, it simply would not do for Jesse Owens, the great track champion, to be seen smoking on network television. Nor did he want any of the young people who idolized him to think that he condoned the use of tobacco. But now that the technicians were coiling their cables and packing their cases, he pulled out a cigarette, lit up, and inhaled. Eventually, this habit would kill him—as it killed Murrow. But he was hooked, of course, and he would just as soon have joined the Communist Party as quit his Camels.

As the crew finally moved his couch and coffee table back where they belonged—into the deep indentations in the carpet—Owens and his wife, Ruth, carefully returned his memorabilia to a display case. A few special items had been freed from the case temporarily, for Murrow and his audience to see clearly. There were the bronzed spikes. And the medals. The laurel wreaths. All the tokens of his youthful greatness. He had collected them nineteen years earlier, in Germany, with the eyes of the world fixed on him, in an atmosphere charged by an ascendant Third Reich, on a continent that would soon convulse in war and genocide.

Nothing Jesse Owens did at the Olympic stadium diminished the horrors to come. He saved no lives. However, for those paying close enough attention, Owens, in Berlin, revealed essential truths. While the western democracies were perfecting the art of appeasement, while much of the rest of the world kowtowed to the Nazis, Owens stood up to them at their own Olympics, refuting their venomous theories with his awesome deeds.

PART
I

1

A Day to Remember

LARRY SNYDER, Ohio State's thirty-eight-year-old track-and-field coach, was worried, running his fingers through his thick blond hair, nervously picking at his face, wondering what could be done to salvage the day, worried more about the health of the best athlete on his team, the best he had ever seen. Jesse Owens was supposed to be warming up for the 100-yard final of the Big Ten championship meet in Ann Arbor, Michigan. Instead, he was moaning in pain.

"Jesse," Snyder said, "look, it's not worth it. If you can't go, you can't go."

"I'll be okay," Owens responded. He tried to smile, to assure Snyder that everything would be fine. But the throbbing in his lower back made it impossible to smile. "It's feeling loose. It's okay," he lied.

Snyder shook his head. He knew Owens well enough to know that he was lying. By this time, in fact, Snyder and Owens were usually able to communicate without speech. In the sixteen months they had known each other, they had become more than tutor and pupil. They had become friends and allies. Theirs was a relationship, too, that seemed less father-and-son and more older brother-and-younger brother.

"No, Jesse, it's not okay," Snyder said. "You don't have to do this."

"Coach, I'm okay. Let me try."

"Fine," Snyder said. "Well, you better get out there."

Unable to lift his shirt over his head, Owens said, "Dave, help me put this on."

David Albritton, Ohio State's gifted high jumper and one of Owens's closest friends — a high school teammate in Cleveland — walked over, pulled Owens's shirt down over his head, and shook *his* head. The night before he had helped Owens climb into a hot bath, where he had soaked for ninety minutes. After a week of baths and rubdowns, Albritton had urged him to rest — not to risk further injury. But Owens insisted that he would compete.

A twenty-one-year-old sophomore, already a father, Owens hobbled onto the track at Ferry Field at the University of Michigan. As usual, an enormous crowd had gathered to see him run and jump. In the mid-1930s, track and field was still a sport of the masses — the top runners and jumpers and throwers were on the same plane as the biggest stars from baseball, football, and boxing. In fact, in 1950, when the Associated Press polled 393 sports writers to determine the greatest athletes of the first half of the twentieth century, track stars led the way. Six finished among the top eighteen — more than from any other sport. On this day, more than 5000 fans were waiting to watch the phenomenon that was Jesse Owens, expecting or at least hoping to see him break some of the world records he had already tied or that already belonged solely to him. For five days it had been raining. But now, on the campus, it was sunny and warm, ideal sprinting weather. A perfect day, too, to come out to see the athlete who was already being called America's great hope for the Olympics the following year in Berlin.

But for Owens, each step was a test, to see if his back could hold his weight. *So stupid*, he thought, *so stupid*. Five days earlier, five days before the biggest meet of the season, he had fallen down a flight of stairs while horsing around with his Alpha Phi

Alpha fraternity brothers in Columbus. The injury was all the more upsetting to Owens because it had seemed that he was getting stronger every week. He was coming off a remarkable meet at Northwestern, where he had set a world record in the 220-yard dash and tied the world record in the 100-yard dash. Still, in the back of his mind, he remembered what his first coach, Charles Riley, had always told him about injuries — that they can establish focus, if they are not entirely debilitating.

Usually Owens warmed up by jogging a quarter mile and stretching. But now he couldn't do either, even though Mel Walker, another Ohio State high jumper, had spent an hour rubbing him down the night before, after the bath. Snyder finally had had enough. He walked over to Owens and put his hands on his shoulders. "Jesse," he said, his eyes searching for some way to gauge Owens's discomfort, "I'm pulling you out of the meet. You can't run."

Without meeting Snyder's gaze, Owens started to protest. "Coach, I wouldn't lie to you. I'll be okay."

"I know," Snyder said, "it's a big meet, a big day. I know. But it's not worth risking all the other days. *You can't run.*"

"Coach, can't we just wait and see how the first race goes?" Now Owens was begging.

Snyder removed his hands from Owens's shoulders. He shook his head. "Fine, you win," he said. He agreed to allow Owens to run the first race.

The day before, Owens had qualified for the final of the 100-yard sprint despite the searing pain in his back. Now all he wanted was a chance to compete in the final. If he couldn't withstand the pain, he wouldn't participate in the three other events he had entered: the 220-yard dash, the 220-yard low hurdles, and the broad jump.

Stiff and angry at himself, he walked to the starting line. Even in his compromised state, he stood out from his competition. The numbers were average: five foot ten, 160 pounds. But the physique was extraordinary. The sportswriters of the time, white and black, often likened him to a big cat or, alternately, to a thoroughbred. His legs were perfectly tapered, his chest barreled and lean, his features sharp. Men as well as women routinely used one word to describe him: beautiful. He was called the Buckeye Bullet for alliterative and aesthetic reasons. And when in motion, he was a sight to see.

As he was limbering up, Owens spotted Riley, the short, frail-looking Irishman who had introduced him to running and jumping in junior high school. Riley had driven all the way from Cleveland, nearly 200 miles, in his ancient Model T to see him compete. If there was one person Owens hated to disappoint, it was Riley — the man he had always considered a second father, the first white man he had met who seemed to be colorblind. When food had been scarce at the Owenses' dinner table, there had always been room for Jesse at Charles Riley's. When Jesse's parents could not afford to buy him running shoes, Riley had dug deep into his meager savings to do so.

Suddenly, as adrenaline started to surge through Owens's veins, the pain subsided. At 3:15 P.M., the cinders crunching under his spikes, he got down on one knee, took the garden trowel that was lying on the dirt, and started digging a hole into which to plant his right foot. Runners still performed this task in 1935, because starting blocks were rarely used.

Now Owens could see both Riley and Snyder. Both wore unfamiliar faces of anxiety, and remarkably, it took Owens a moment to remember why they were anxious. Oh, yes, the pain. But for him, it had disappeared. At the moment he was about to run, he had always been able to shut out all distractions. His ability to focus — for all sprinters an invaluable asset, because their margin for error is so slim — was one of his great strengths. Settling into his starting position, Owens felt not soreness but the almost indescribable elation he always felt when he was about to run at full speed. At a certain level, he knew that when he ran at full speed, he was capable of running faster than any human had ever run. But the joy came from a deeper place. The knowledge of his own capacity for greatness made him thankful, humble, and proud.

Getting up from the dirt, stretching his arms, then his hamstrings, then, again, his lower back, Owens thought to himself, *Pain? What pain?*

One hundred yards away, six timers were standing by the finish line, stopwatches at the ready. Like the runners, they were waiting nervously for the gun. When it was fired, Owens was off, but as usual, not very quickly. The greatest sprinter the world had ever known was never a fast starter. But he could afford to wait. "About everybody beat him off the start," said Bob Collier of Indiana University, who was also in that race. "But after thirty yards, it was no contest."

As was his custom, 30 yards from the starting line, Owens exploded, like a powerful engine that has finally warmed up. His short, effortless strides attained a spectacular rhythm, his body accelerating without any apparent effort. His famous, textbook form held steady — his posture never altered, no motion was wasted, his upper body was perfectly straight. If his back was still hurting him, no one could tell. Ten yards from the tape, he was still accelerating, a blur. As he broke the tape — his eyes straight ahead, because he was confident that no one was anywhere near him — the stopwatches clicked and a wave of applause rose from the crowd.

Three timers had clocked Owens at 9.4 seconds, the other three at 9.3. Unfortunately, the three official timers — three were alternates — clocked him at 9.3, 9.3, and 9.4, and as was customary, the slowest time would be the official time. Big Ten timers, including the head timer, Professor Phillip Diamond of the University of Michigan, were known for their slow thumbs and stodgy rules. Unlike most timers, Big Ten timers held that a runner crossed the finish line only with his so-called center of gravity, not with his nose, arm, or knee. In a further attempt to prevent his timers from being too generous, Diamond advised them to click their stopwatches only when they saw a runner's back foot cross the finish line. His fussiness probably cost Owens a world record; as it was, his 9.4 tied the world record he already shared with the two-time Olympic gold medalist Frank Wykoff.

In the stands, Riley was still on his feet, still cheering. On the track, Snyder, always emotional, was jumping up and down, pumping his fist. He had been worried that Owens might throw out his back; now he ran to the spot where the sprinter was catching his breath to offer his congratulations. "Jesse," he said, "that was phenomenal. I don't know how you did it."

"I could have gone faster," Owens replied flatly. Then, stretching out his back, he turned to answer a reporter. "I'm not bragging," he said, "but I really did get a bad start. Frankly, I am a little bit disappointed."

But there was no time for him to linger in his disappointment. Fifteen minutes after the 100-yard dash, he was preparing to launch his body into the broad-jump pit. In the previous day's *Los Angeles Times*, Snyder had predicted that Owens would soon break the world record, then held by Chuhei Nambu of Japan, of 26 feet, 2 inches. Over the course of more than two thousand years — the ancient Greeks were broad-jump enthusiasts — the record had moved barely 2 feet. In the *Times*, Snyder told Francis J. Powers that Owens "can do a full 27 feet if he will concentrate on his takeoff." For his part, Powers ventured a guess: "It may be that Owens will find the board in the Big Ten meet in Ann Arbor, and, if so, Nambu . . . and others who have planted their spikes on the records will be dismissed from the books."

Now, as he readied himself to jump, Owens, in a rare display of showmanship — Riley had trained him to accept victory and defeat (however rare that might be) with the same good grace — gently placed a white handkerchief 26 feet, 2 inches from the takeoff board. He wanted a target — Nambu's mark — even if it was barely discernible from the takeoff board. Like Snyder, Owens knew that the broad-jump record would eventually be his. In training sessions he routinely jumped farther than Nambu had jumped; it was just a question of when he would get all the details right in a competition. For Owens, like most sprinters who moonlight as long jumpers, jumping was much more punishing than running. For this reason, and because he was so much better than almost anyone he competed against, he rarely jumped more than once or twice in any given meet. Those jumps, though, were usually awe-inspiring.

As Owens was preparing to hurl himself down the track and toward the pit, Snyder came up to him and told him, needlessly, to concentrate on his form, to generate as much velocity as possible before he hit the board — and, most important, to measure his steps carefully so that he would neither foul nor leap too far back on the board. Form didn't matter much to Owens, because he always had it.

At 3:35, as Riley and Snyder — and everyone else at Ferry Field — trained their eyes on him, Owens started running toward the takeoff board, 108 feet away. As the distance dwindled, his speed rapidly increased. He was moving fluidly, his arms pumping, his gait natural. He hit the board in full stride — there was no stutter-stepping to prevent fouling. Then he was in the air — for less than a second, but his body got so high so fast that it seemed he might sail entirely clear of the pit. The crowd knew somehow that they were witnessing that rarest of feats, unequaled athleticism matched with flawless execution, and gasped. Michigan State's athletic director, Ralph Young, was standing beside the pit, which was lined with dozens of spectators. Young later told Larry Snyder that Owens jumped over his head "by about a foot," and a famous photograph seemed to prove his assertion. Finally Owens's feet crashed into the sand. He knew too. He had flown directly over the handkerchief.

Still, it took a few moments to measure off the distance. Then the track announcer, Ted Canty, turned on his megaphone.

"We have a new world's record," he said, stating the obvious. "Twenty-six feet, eight and one-quarter inches!" Nursing a sore back, Owens had broken the world record by more than half a foot.

Albritton and Walker rushed over to congratulate him. Riley pumped his tiny, clenched fist. Snyder smiled, as if to say, "I knew it. It was only a matter of time."

This time Owens was tempted to jump again, just to see if he could reach the 27-foot mark. But there was no time. He was due at the starting line for the 220-yard dash. The world record, set by Roland "Gipper" Locke of the University of Nebraska in 1924 and tied by Ralph Metcalfe in 1933, was 20.6 seconds. Standing on the field not far from the track, seventeen-year-old Tom Harmon had gotten caught up in the excitement. Harmon was in Ann Arbor on a recruiting trip — the Michigan football team wanted him badly — and he would go on to win the 1940 Heisman Trophy for the Wolverines. But on this day he was simply an awed teenager. Owens, Harmon thought, was "absolutely beautiful."

At 3:45 — just ten minutes after hurtling nearly 27 feet through the air — Owens, who had now completely forgotten how sore his back had been, threw himself down the track once more. This time his start was clean, and for all but the first few yards of the race he was utterly alone. Even without anyone to push his pace, he picked up speed with each stride, until it seemed that his feet were barely skimming the track. Then the stopwatches clicked again. Another world record — 20.3 seconds. No one had ever before set two world records on the same day. In the stands, Charles Riley's was only one among thousands of voices going hoarse in Owens-induced ecstasy.

Snyder, too, was truly pleased. In two years at Ohio State, Owens had been spectacular, but barely more spectacular than he had been in high school. Snyder's fellow coaches had not given him any credit for the records Owens had set; on the contrary, they thought — most of them, anyway — that he had either stalled Owens's progress or that Owens had nothing more to

give. "Every coach in the Big Ten was watching me with a critical eye," Snyder said, "to see how I would handle him." Now no
one could deny Snyder's role in the development of Jesse Owens. Owens had come to Snyder gifted but raw; now he was
gifted and invincible. The victories belonged to Owens, but finally there would be acclaim for Snyder too. More than ever,
their reputations were linked.

The Jesse Owens whom Snyder was now watching was unlike any track-and-field athlete he or anyone else had ever seen.
Owens was graceful and swift and unflappable. Soon sportswriters around the United States would take to calling him, among
other things, "the Ebony Antelope." The predominant track star
of the previous decade, Paavo Nurmi, the Flying Finn, a middle-
and long-distance runner, wasn't beautiful to watch. You can
admire the stamina of a great distance runner, but you can't really enjoy watching someone run 10,000 meters, 25 laps around
a track. Owens, in contrast, brought joy to the huge crowds that
gathered when he competed. His races became events.

In the 1930s, track and field didn't attract quite as many fans
as baseball and boxing and horseracing, but it was far from a
niche sport. The results of the big meets were printed in daily
newspapers of every size. The big national columnists — Grantland Rice and Paul Gallico and others — dedicated hundreds
of column inches every year to runners, jumpers, and weight
throwers. Jim Thorpe, the 1912 Olympic decathlon champion,
was widely considered the greatest athlete who'd ever lived.
(These days, no one thinks that the Olympic decathlon and
pentathlon champion is the world's greatest athlete. He's considered little more than an acutely focused specialist, someone who's spent years mastering the intricacies of the javelin
and pole vault but probably couldn't play tight end in the National Football League.) Tens of thousands of people regularly
filled stadiums from Los Angeles to Philadelphia to watch college track meets. Young men with running ability weren't necessarily snatched up by football coaches to play running back and

wide receiver; most of them went out for the track team first and then maybe the football team, if their track coaches allowed it.

The final event in which Owens competed that day was the 220-yard hurdles. For all his gracefulness, he was an awkward as well as a reluctant hurdler. His form was nonexistent. He jumped too high and never in stride. Additionally, he didn't much like the hurdles. He didn't like anything that slowed him down. But as long as he was not feeling too much pain, Ohio State — and Larry Snyder — needed him to run the hurdles. At 4 P.M., having set two world records and equaled another in the previous forty-five minutes, Jesse Owens was once again ready to race. As he crouched into position, a sharp pain shot up his back — but only for an instant. It was gone by the time he had settled into position — a reassurance, actually, that his injury was manageable. Then, the gun.

Owens ran hard, then, reaching the first hurdle, slowed considerably, cleared it, then ran hard again, slowed, cleared another hurdle, then accelerated, again and again and again. At each of the ten hurdles, the field gained on him; on the flats in between, it faded. "It was like an accordion," Francis Cretzmeyer of Iowa, who finished third, said. Still, Owens won by 5 yards. His time, 22.6 seconds, beat the world record by four tenths of a second. But the world's best 220-yard hurdler had so little regard for the event that he rarely chose to compete in it.

In the stands, when the time was announced, Charles Riley shook his head slowly, marveling at what he had witnessed. The rest of the spectators were less pensive, cheering madly, rushing the field, clamoring to get near the young man who had just set his third world record in less than an hour. (Technically, Owens had set two more world records. His times in the 220-yard dash and 220-yard hurdles were also recognized as the world records for the 200-meter dash and 200-meter hurdles, respectively, because, despite the fact that the metric distance is about four feet shorter than the imperial distance, he ran the imperial events faster than anyone had ever run the metric events.) No one at Ferry Field that day could have known that Owens's records would all prove remarkably durable, especially the broad-jump record. It would stand for twenty-five years, an eon in track and field.

At the recent summer Olympics, in Athens in 2004, sixty-nine years after Owens's historic day in Michigan, his 26-foot, 8¼-inch leap would have been good for ninth place. Imagine if Owens had had the advantages the athletes in 2004 had had — the equipment, the supplements, the rubberized track, the wind-resistant clothing. His records might still stand.

"I broke out that Saturday in Michigan," Owens later recalled. "It wasn't only overcoming physical pain; it had to do with overcoming something psychological inside myself." He could not have known that there were still greater moments ahead. Physically, perhaps, he peaked at Ferry Field. But in every other way he was still improving.

That afternoon, Alvin Silverman of the *Cleveland Plain Dealer* managed to track down Owens for a phone interview. To Silverman, whom he could barely hear, he repeated what he had said earlier. "I want to tell you something, if you won't think I'm getting swell-headed. I really believe I can run the 100-yard dash in 9.3 seconds. I'm not bragging, but I really did get a bad start."

Silverman then asked Owens how Charles Riley had reacted to his feats.

"Mr. Riley? He didn't say anything to me. He started to give me the big rah-rah, and then I'll be doggone if he didn't break down and start to cry. He put his head on my shoulder and cried like a kid."

The crowd that gathered outside the locker room was so thick that Owens decided he had to escape through the back. Snyder said farewell to him, then he hopped out the bathroom window and into Charles Riley's Model T. Together they would ride back to Cleveland — and a hero's welcome.

2

Out of Alabama

T HE LEGS. They were what Charles Riley first noticed. They were perfect. Before he knew the boy's name, before he knew whether he could run, before anything else, there were those legs. "My father's long, lion-spring legs," Jesse Owens called them.

A fifty-year-old coach and physical education instructor at Fairmount Junior High School in Cleveland, Riley had observed thousands of boys and girls. But he had never seen legs like these. Even when Jesse Owens was just walking, his oddly symmetrical and powerful legs made it seem as if he were moving more swiftly than everyone else. Immediately Riley knew there was something special about this Owens kid.

Born on December 11, 1878, Riley had grown up in Mauch Chunk, Pennsylvania, and as a young man had worked in its mines. At the turn of the twentieth century, Mauch Chunk was one of America's richest cities — the so-called Switzerland of the United States — with more millionaires per capita than anywhere else. But the Rileys were no millionaires. (Oddly, Mauch Chunk is the only town in the United States deeply associated with America's two greatest track-and-field athletes — probably the two greatest track-and-field athletes anywhere, ever. In 1954, in an effort to attract tourists as its economy stagnated, Mauch Chunk and neighboring East Mauch Chunk agreed to merge and become Jim Thorpe, Pennsylvania, despite the fact that Thorpe had never been to either Mauch Chunk or East Mauch Chunk.) Riley was slight, pale, and nearsighted, but physical exertion thrilled him. When he saw Jesse's legs, he recognized in them limitless potential.

For about a year Riley simply watched and waited. When Jesse was running and jumping at recess or in his gym classes — naturally, no one could catch him — Riley was there, taking mental notes, planning the boy's future without the boy's knowledge. Then finally, one day when Jesse was thirteen, Riley decided it was time to introduce himself.

"Your name's Jesse, am I right?" he said.

In Alabama, where Jesse had been born and had spent the first nine years of his life, it had been rare for him to be addressed by white men. In Ohio, that had changed. Still, he was too nervous to speak. He could only nod.

"How would you like to be on the track team when you get into high school?"

"I'd like that plenty," Jesse said. He had noticed the coach watching him. He had hoped that he would make him an offer.

"Well, then," Riley said, "you'll have to do more than we do in gym class. Are you willing?"

"Yes, sir," Jesse said.

Riley explained that if Jesse agreed, he would spend about ninety minutes every day after school learning the finer points of running. Unaccustomed to the attention of adults other than his parents, Jesse was thrilled to have someone paying so much attention to him and simply kept nodding.

"Well, then, see you tomorrow, Jesse," Riley said, and walked away.

In his eagerness to run, Jesse had forgotten that he could not train after school. He had a job delivering groceries and another working in a greenhouse. He didn't work just to have pocket change. These were jobs that helped feed his family. Now, in an instant, his elation turned to anxiety. In his thirteen-year-old mind, he could see his whole life falling apart because of a sched-uling conflict. He chased after Riley, caught him — of course — and laid out his dilemma.

Riley said, "That's no problem. You'll run before school, won't you?"

"Yes, sir," Jesse said.

Charles Riley had five children of his own, three daughters and two sons. One son was not interested in sports, the other disabled. Naturally he pinned his sporting hopes on his stu-dents — and none ever received more of his attention than Jesse Owens.

At first Jesse felt guilty spending so much time practicing a skill that seemed to have no practical application. He needed to work to help his family, and all the time he spent running he could have been spending at work. But he loved running be-cause he knew he was good at it, and Riley had rearranged his hours to make it possible for Jesse to train with him. Together they spent hours in the school gymnasium, Jesse harnessed to the wall, running in place, perfecting his form. Riley had a monkish devotion to the value of repetition. It wasn't enough for him that Jesse was incredibly gifted. He wanted to harness the boy's gift, hone it, perfect it. Sensing this, Jesse never rebelled against Riley's increasing demands. Without Riley's ever having to say it, Jesse knew the coach was only doing what was best for him. As he exhausted himself in the gymnasium or running on the muddy track around the football field, Riley was right there with him, his eyes focusing on his stopwatch and then on Jesse's form. "Knees up! Head up! Watch your form!" Riley would say. He would not allow Jesse to drag his feet or arch his back. He would not tolerate bad habits. He would not let their gift go to waste.

It was an odd feeling for Jesse to have someone take so much interest in him. His mother had always told him that he was spe-cial, but that was his mother. Now, having that idea reinforced by an authority figure — a white man, no less — he blossomed. With each blistering lap on the track, with each pat on the back from Riley, Jesse allowed himself to believe that there was a future for him in running, in simply doing what felt as natural to him as breathing. With Riley pushing him, he was empowered, and for a black teenager in the 1920s, that in itself was an achievement.

Like most great heroes, Jesse Owens had his own creation myth — which is not to suggest that it was untrue. By his own reckoning, he was more or less molded by the tragic events and desperate circumstances of his childhood in Alabama and then Ohio. Born on September 12, 1913, in Oakville, Alabama, James Cleveland Owens was the tenth and last child of Henry and Mary Emma Owens. He sometimes said later in life that his early childhood in Alabama was essentially happy, because he had no idea how poor he was. But there was nothing genteel about the Owenses' brand of poverty. For sharecroppers like Henry Owens, every day was filled with struggle. If there was too much rain or not enough, if there was too much frost, a crop could fail — and then it would take all his resourcefulness merely to feed his family. For the Owenses, everything but food, shelter, and the simplest clothes was a luxury that simply could not be afforded — even medical care.

For Jesse Owens, the defining moment of his youth — the story he told over and over — revolved around a fibrous bump he noticed on his chest the day after he turned five. At first he thought it would just go away. But within a few days he could see it growing and feel it pressing against his lungs. Eventually, J.C., as his parents called him, could no longer bear the discomfort. He told his mother. Not always the most reliable narrator of his own life story, Owens later reconstructed the subsequent conversation he claimed he overheard between Henry and Emma Owens:

"We've got to do something," Emma said to Henry.
"You took one off his leg once, Emma."

"But this one's so big. And near his heart."

"Emma —"

"Don't say it, Henry!"

"I'm going to say it. If the Lord wants him —"

All these years later, it is impossible to say who had the greater talent for melodrama, Henry Owens, in his fatalism, or Jesse Owens, in his storytelling. Certainly it is possible that Jesse misrepresented Henry's exact words, but undoubtedly Henry was a man who had come to expect the worst in any situation.

The son of former slaves, Henry Owens had grown up in Oakville, not far from Georgia but a world away from the nearest big town, Decatur, which is 20 miles to the northeast. He spent most of his life scratching out a living as a sharecropper. By most accounts, he lived in mortal fear of his landlords — and other white men. For southern black men of his generation, deference to white men was nothing less than a survival imperative. Between 1882 and 1902 there were more than one hundred lynchings each year in the United States, the vast majority perpetrated in the South. All his life, Henry Owens avoided making eye contact with whites.

Emma Owens was as ambitious as her husband was timid. Against all odds, she held out hope that her children's lives would be less bleak than hers. She had grown up in less desperate circumstances than her husband. She knew that there was a world outside Oakville, even if Henry didn't. From the beginning, Jesse took after his mother, not the father who had long ago learned to keep his expectations low — even when the subject was his son's health.

A few nights after little J.C. brought the growth on his chest to his parents' attention, he was lying in bed. His mother came to him and said, "I'm going to take the bump off now, J.C." The Owens family could not afford the services of a physician.

As Henry Owens wept quietly in the corner, Emma Owens sterilized a kitchen knife over a flame. Then she started cutting into her son's chest. J.C. bit down hard on a leather strap. No sound penetrated the strap, but tears flowed down his cheeks. Dark, thick blood started pouring out of him. Emma moved the knife around the edges of the lump, searching for its contours, trying to determine its size and consistency. It was bigger than she had thought. It seemed to her as if the blade was inching too close to J.C.'s heart. Still, she kept cutting. Finally it was done. She extracted the gelatinous lump, and in its place was a hole the size of a golf ball, not oozing but spurting blood. Emma tried to stop the bleeding, but for days it continued.

Three nights after the makeshift operation, J.C. rose from his bed and walked to the front door of the tiny house. He later said that he could hear his father praying outside. "Oh, Lord Jesus," Henry said, "Please, please, hear me. I know you hear everything, but this saving means everything. She'll die if he dies — and if she dies, Lord, we'll all die — all of us."

As he listened to his father's desperate prayer, J.C. was getting weaker. "My body was emptying of blood," he later wrote.

"He's my last boy," Henry continued, still on his knees. "J.C.'s the one you gave me last to carry my name. She'll die if you take him from me. She always said he was born special."

Then J.C. walked into the night, into his father's arms. "Pray, J.C.," Henry Owens said. "Pray, James Cleveland." Together, father and son knelt and prayed.

Within minutes, Jesse Owens later wrote, the bleeding stopped. The growth was probably a fibrous tumor.

As an adult, Owens usually described his hardscrabble youth in Oakville as beyond miserable — an endless cycle of poverty, hunger, and humiliation. But on other occasions he said that he had been happy in Alabama. Knowing nothing of the world beyond Lawrence County, he never considered himself deprived. "We never had any problems," he said. "We always ate. The fact that we didn't have steak? Who had steak?"

J.C. and his six brothers and three sisters were forced to spend about one week each year picking cotton, but most of the time they were free to play out in the fields that Henry Owens farmed.

Religion was a constant for the Owens family. They were devout Baptists, regular congregants at the Oakville Missionary Baptist Church, where Henry was a deacon. During the week it was the schoolhouse for Oakville's black children. J.C., though, like most of his siblings, wasn't there enough to learn much more than how to read and write. More often he was out in the fields, running barefoot, running because he loved it and because there was little else to do.

"I always loved running," he said about his youth in Oakville. "I wasn't very good at it, but I loved it because it was something you could do all by yourself, all under your own power. You could go in any direction, fast or slow as you wanted, fighting the wind if you felt like it, seeking out new sights just on the strength of your feet and the courage of your lungs."

In 1922, when Jesse was nine years old, Emma Owens again took the initiative and made a decision that forever changed the family's fortunes. Her daughter Lillie had already joined the vast migration of southern blacks to the industrial North. Like thousands of blacks from Alabama, Lillie had moved to Cleveland, which had a reputation (not entirely deserved) for racial progressiveness. In Cleveland, Lillie had thrived. She wrote to her mother, urging the entire family to say farewell to its dead-end existence in Oakville. Emma was instantly eager. She refused to settle for what life offered in Alabama — little more than subsistence and the comfort of familiarity. But Henry Owens was, understandably, scared.

"It's crazy to go on like this, Henry," Emma said to her husband one day. J.C. was again within earshot.

"We've got no choice," Henry said. Jesse later said it was the only time he heard his father raise his voice.

"Folks have done it," Emma countered, meaning not just Lillie.

"Who?"

"You heard about it, same as I have. Folks have done it!"

"And never been heard from again!"

"We're nearly starving here," Emma said, her tone softening.

"It can't be done, Emma. So we're not going to do it. That's my final word."

Eventually, Emma got her way. Soon enough, the Owens family was packing up for Cleveland. Jesse said his father was so nervous that he couldn't stop shaking. When it became clear to Jesse that the family really was leaving Oakville, he said to his mother, "But where're we gonna go, Momma?"

"We're going on a train," she said.

"And where's the train gonna take us, Momma?"

"It's gonna take us to a better life."

In Cleveland, J.C. enrolled at Bolton Elementary School. It was there that J.C. became Jesse. One day in class, a teacher asked him his name. Speaking in his Alabama twang, J.C.'s reply sounded to the teacher like "Jesse."

"Jesse?" she said, just to be sure.

"Yes, ma'am, Jesse Owens," J.C. said. He was in no position to argue the point.

Life in Cleveland *was* better for everyone in the Owens family, except Henry Owens. Even as his wife and children found work, Henry struggled to get a good-paying job. He was forty-five and had known nothing but sharecropping. He found work in a steel factory, but it was irregular. He was too limited. Outside Oakville, he simply wasn't comfortable in his own skin, and that was apparent to everyone. Emasculated, in his own mind, by his inability to support his family, he grew increasingly bitter. There was so little he could offer his youngest son, who, despite his father's gloominess, somehow managed to maintain his innate enthusiasm for life, enthusiasm he soon focused on running under Charles Riley's approving eye.

One day when Jesse was fifteen, a 100-yard race was arranged on the sidewalk on East 107th Street. Riley had his stopwatch out. He was measuring Jesse less against his opponents than against time. As usual, he wanted to measure his student's progress.

"All right, boys," Riley shouted from the finish line. "On your marks, ready, set, go!"

In his leather-soled tennis shoes — a gift from Riley — Jesse flew down the pavement, far outpacing his schoolmates, pulling up only when he reached his coach. He had wanted to impress Riley, and he was breathing hard because it was hard work trying to outdo yourself when you were Jesse Owens. The coach was not surprised that Jesse had won easily. But he was surprised when he looked at his stopwatch.

"It can't be," he said. "I must have punched it too soon."

"How fast, Mr. Riley?" Jesse asked, still panting.

"It can't be," Riley repeated.

"How fast?" Jesse was now curious.

"Eleven flat," Riley said, removing his glasses as if they had just lied to him. "Eleven flat."

"Is that fast?"

"Too fast," Riley said, shaking his head. "No eighth-grader can run that fast."

Eleven flat. In other words, at fifteen years old, in cheap shoes and street clothes, Jesse Owens was more than just a strong local talent. He was world-class — and not just for his age. The world record was 9.6 seconds — held by eight sprinters, all adults, who had set it in spikes, on real tracks, with real competition.

To confirm his findings, Riley had Jesse run the same distance the following day. Again he timed him at 11.0 seconds, which proved that Jesse was one of the faster people on the planet, regardless of age.

Soon after his sprint down East 107th Street, Jesse ran his first official race. It was a quarter-mile — not his best distance, but one he had run many times training for the sprints. Unsurprisingly, he ran the race like a sprint, dashing to the front of the field, only to be overtaken, 50 yards from the finish line, by two of his competitors. At the tape another runner overtook him, so he finished fourth.

Embarrassed, Jesse could not face Riley for almost thirty minutes. Finally he approached his mentor and said, "I thought I'd win, Mr. Riley. I should have. Why didn't I?"

"Because," Riley said, "you tried to stare them down instead of run them down."

"I don't get it," Jesse said. "What do you mean?"

"I'm not going to tell you," Riley said. "I'm going to show you. Do you work Sunday afternoons too?"

"No, just in the morning, and then I can do what I want the whole day."

"Well, I'll pick you up in front of your house at one o'clock."

"Where are we going?" Jesse asked. His T-shirt was still wet, but he'd long since stopped sweating.

"We're going to watch the best runners in the whole world. I'll pick you up at one."

By this time the Rileys had become Jesse's second family. He dined with them often and was so frequently in his coach's company that he was kidded about his "other" father. In his melancholia, Henry Owens did not seem to notice that his son was never around much. Emma was pleased simply to know that Jesse was being treated well. Riley had told her that Jesse was capable of greatness, confirming what she had always said about her baby boy. So there was nothing unusual about Jesse's spending a Sunday with Charles Riley — it was just another day on which he would be trained and encouraged.

For his part, Jesse waited eagerly for Riley to pick him up. Of course, he was expecting that they would be driving to see Charley Paddock, the great Olympic star, or Eddie Tolan, another Olympian. He was expecting a lesson in form from one of the men he aspired to emulate. Instead Riley took Jesse straight to the track — the racetrack. He wanted him to see the thoroughbreds. Riley wasn't a gambler — nor was he particularly fond of the seedy scene at the track — but he had an Irishman's passion for beautiful horses. He loved watching them run, their muscles rippling under their shiny coats, the flare of their nostrils as they reached top speed. To Riley, horses were the purest runners on earth, unburdened by human flaws such as vanity and egotism. He hated runners who showed up their opponents with their body language and facial expressions. He hated showmanship.

Riley walked Jesse straight up to the rail, where together they spent the entire afternoon just watching the thoroughbreds run race after race. "Don't talk, Jesse," he said. "Watch."

And Jesse watched. He watched the horses as they galloped effortlessly down the stretch. He too couldn't help but admire their form, the way it seemed that all their energy was expended but none wasted.

Finally, after the tenth and final race, Riley turned to the boy and asked, "Well, what did you learn today?"

"Well," Jesse said, his hand still clasping the rail, "the way they move — the legs and the whole bodies of the horses that can get the lead and keep it — is like they're not trying. Like it's easy. But you know they are trying."

"And what about their faces?" Riley asked, getting to his primary point.

"I didn't see anything on their faces."

"That's right, Jesse," Riley said, lifting his bony finger to Jesse's chest. It was an accusation. "Horses are honest. No animal has ever told a lie. No horse has ever tried to stare another one down. That's for actors. And that's what you were doing the other day. Acting. Trying to stare down the other runners. Putting your energy into a determined look on your face instead of putting it into your running."

Soon thereafter Jesse found himself racing some local teenagers in a 220-yard dash. He went all out in the first 100 yards; then, just when he expected to find himself tiring, he realized that he still had more to give, more than he had ever known he had. He kept pushing, and pushing harder, and then, to his surprise, he was overtaken. He had pushed himself too hard. Still, though, he kept pushing, fighting to catch up. He nearly did, but not quite. Frustrated, he kept running after the race was over, until he passed the two teenagers who had outsprinted him.

"Congratulations, Jesse," Riley said. His face was somewhere between a smirk and a smile.

Jesse, panting, thought Riley was making fun of him.

"You think you lost today, but you're wrong. You won. Even when the race was over, you didn't stop. You won — and you overcame your greatest opponent."

Jesse knew that he meant himself.

From that point on, as Owens would later recall, he rarely lost a race. Eventually his style would be called effortless, emotionless. He never appeared to strain, and his face rarely betrayed his thoughts or feelings, whatever they may have been. He ran the way Charles Riley taught him to run — like a thoroughbred.

In 1928, when Jesse was still fifteen, Riley arranged for Charley Paddock to speak at Fairmount. After his speech, Paddock was introduced to the young man who worshiped him. Jesse knew all about Paddock because Paddock had been the first great sprinter to alchemize Olympic gold into international fame. He was living proof of the worthiness of Jesse's aspirations. In 1920, at the Olympic games in Antwerp, Belgium, Paddock had won the gold medal in the 100-meter dash and the silver medal in the 200-meter dash. Four years later, in Paris, he won a gold medal in the 4 x 100-meter relay and again won the silver medal at 200 meters.

Like so many young men momentarily in the presence of a significant figure, Jesse was profoundly influenced by his brief encounter with Paddock. Here in the flesh was the man whose records he sought. Paddock had an aura; Jesse could sense it, and he wanted one too. It mattered not a whit to Jesse that Paddock was a white man, to whom, unlike Jesse, no doors were barred at that time.

In the fall of 1930, just as he was turning seventeen, Jesse finally entered Cleveland's East Technical High School. He and Riley were not parted, however. Edgar Weil, East Tech's novice track coach, asked Riley to assist him. Riley, of course, eagerly assented to Weil's request. For the next three years Riley continued to develop Jesse's form, as Jesse Owens became a world-class sprinter and broad jumper. By the spring of 1932, when Jesse was eighteen, he had progressed to the point where it was not unreasonable for him to think he might qualify for the American Olympic team that would be competing in Los Angeles that summer. In fact, on June 11, in Cleveland, he ran 100 meters in 10.3 seconds, breaking the world record Paddock had established in 1921. Meet officials, though, did not allow the record, because the tailwind was too strong. Still, Jesse's performance qualified him for the semi-final central Olympic track-and-field tryouts at Northwestern University in Evanston, Illinois, the following week.

In Evanston, Jesse Owens was exposed for the first time to world-class competition on a grand scale. Three hundred of the best track men in the Midwest and a few from the South assembled at Dyche Stadium, including Tolan and the hurdler Glenn Hardin. At eighteen, Owens was younger than most of his rivals but hardly a child. That day, though, for perhaps the only time in his life, he was out of his league. Tolan, the first black man to win two Olympic gold medals, was twenty-three. It is unlikely that he had ever paid any attention to the exploits of East Tech's Jesse Owens.

Racing against America's best that day in Evanston, Owens failed to make much of an impact. He did not qualify for the Olympic trials in Palo Alto. He could not defeat Tolan. This time, he could not emulate the thoroughbreds Riley had shown him. Still, he could not help staring at the men he was racing against, the only men he had ever encountered who were so clearly faster than him. At the same time, he saw something that encouraged him. Like most great winners, Owens, even before his accomplishments caught up with his gifts, was blessed with remarkable self-awareness. He was analytical, even if at this time he did not quite know how to communicate to anyone else what he was thinking. That day in Evanston, he looked closely at Tolan and James Johnson and George Simpson and saw that they were human. He saw flaws in their styles. He recognized weaknesses that Riley had pointed out in other runners. Most important, he saw that a day would come when he would be able to compete against them, indeed, a day when he would defeat them. As he gathered his few belongings and walked off the field, he could not help noticing all the people who had crowded around Tolan, who had, as expected, won the 100-meter dash. *One day,* he thought, *the crowds will gather around me. One day, after I gain a little weight and grow an inch or two.*

While Owens weighed barely 140 pounds, Tolan was thickly muscled. Owens understood that with a little more power, he would rise to his level. And even as he lost — a strange, unpleasant experience — he could sense that it was all just a matter of timing. His innate optimism and his burgeoning self-confidence helped him realize that in four years, when *he* was twenty-two, he would be the sprinter against whom all the others would be measuring themselves. All it would take was dedication. It was clear — to him, and to everyone else — that he had the talent.

Still, he was disappointed — and embarrassed. When Alvin Silverman of the *Plain Dealer* reached him after the meet, Owens said, "I haven't got the heart to see Mr. Riley. He must be terribly ashamed of me. I don't know what was the matter. I ran as fast as I could, and Lord knows, I tried. But I just didn't have it. I'm going to work my heart out from now on. I betcha." Owens even felt that he had let Silverman down. "It was nice for you to try and make me feel good," he told the reporter, who had followed his high school career closely, "but I'll bet you're as much ashamed of me as anybody else."

One issue had been weighing heavily on Owens as he tried to keep up with Tolan and Johnson. His girlfriend, a lively, pretty girl named Minnie Ruth Solomon, was eight months pregnant. Ruth, as she was called, was only sixteen. In Cleveland's black community, her pregnancy was not quite scandalous, as it would have been in white society. Still, it was not exactly something either family was thrilled about. But both families had for years assumed that Jesse and Ruth would eventually build a family of their own. "I fell in love with her some the first time we ever talked," Owens later said.

What he liked most about Ruth was the way she carried herself — with dignity, like a princess, which is how he treated her. Her family had come from Georgia, and though the Solomons were poor, as poor as the Owenses, Ruth, well dressed and immaculate, seemed untouched by poverty. When they first met, in junior high school, Jesse carried her books and asked her to marry him. She told him yes, but said they would have to wait. In fact, even as Ruth's delivery date approached, there were no plans for a wedding. Nevertheless, it was assumed that they would stay together and that Jesse would find some way to help support his child — although it was clear that the child would be supported mostly by Ruth's parents.

On August 8, 1932 — five weeks after the regional Olympic trials — Ruth gave birth to Gloria Shirley Owens. In later years, Jesse would frequently claim that he and Ruth had married a few weeks before Gloria's birth. But no contemporaneous records exist anywhere showing that they married in 1932.

Thirteen days after becoming a father, Owens was back on the track, at an invitational meet in Cleveland that included several elite European runners who were on their way from the just-completed Olympics to the East Coast. More than 50,000 people crammed Municipal Stadium to see the stars from the games — and their own Jesse Owens. Fatherhood apparently agreed with him. He ran 100 yards, on a curved track, in 9.6 seconds, shattering the field, which was reduced to reverence. Erich Borchmeyer of Germany, a silver medalist in Los Angeles in the

4 x 100-meter relay, and Gabriel Salviati of Italy, a bronze medalist in the same event, "hunted up interpreters to extend lavish praises."

For the Olympians, it was time to head home. For Jesse Owens, it was time to head back to East Technical High School for his senior year. His athletic achievements had made headlines across the country, and now coaches interested in procuring his services would be swooping down to claim him for their teams. In the black press in particular, the subject of Owens's choice of college was an important story. It was hoped by some black journalists that the athlete would make a political statement by choosing a progressive school, a place where blacks had been made to feel relatively comfortable. Even the far-off *Chicago Defender* entered the debate. "He will be an asset to any school," a *Defender* editorialist wrote, "so why help advertise an institution that majors in prejudice?"

But Ohio State — far from a bastion of progressiveness — was the power in collegiate track and field in Ohio, and Owens was an Ohioan. It also didn't hurt that in Larry Snyder the Buckeyes had an eager and innovative coach who met with Charles Riley's approval.

In just one year as the head coach in Columbus, Snyder had become well known for some of his quirky training theories. It was considered highly unusual, for instance, that he had his athletes train to the strains of music being played on the phonograph. Snyder said the musical rhythm helped his runners relax and find their own personal rhythms. He also made a point of converting his middle- and long-distance runners to the heel-and-toe style of running, which had been so expertly practiced by Paavo Nurmi and the other Flying Finns of the 1920s. American distance runners had always run more or less flat-footed.

Snyder did not personally recruit Owens. That task was handled by some boosters in Cleveland. Like everyone else in track, though, Snyder was watching closely when, on June 17, 1933, in Chicago, his future charge delivered on all the promise of his talents. In a performance that presaged the miracle day at Ferry Field, Owens equaled Frank Wykoff's 9.4-second world record at 100 yards, despite stumbling out of the blocks, and set scholastic records in the 220-yard dash and broad jump. But Snyder knew Owens only by reputation when they first met, in early 1934. Like all freshmen at the time, Owens was barred from intercollegiate competition. Nevertheless, on the first day of practice for the spring season, he reported to Snyder's office, ready to run.

"Coach, I'm Jesse Owens." He was, typically, immaculate. His sweatshirt, his shorts — they both looked like they had been pressed by Emma.

Snyder looked him up and down. "The great Jesse Owens." He paused, admiring his legs. "I want to know all about you. First, let's get you out on that track."

They worked together that day, just getting acquainted, not knowing, of course, where their partnership would take them. But the bond was immediate. They liked each other — and soon enough respected each other.

Eventually, inevitably, Snyder would come to be remembered not for the music or the heel-and-toeing. Instead he would be remembered for nurturing the skinny kid from Cleveland, the one who would become the greatest track star ever.

3

Vincible

I N THE WAKE of Owens's big day in Ann Arbor, no less a personage than Will Rogers was overcome with enthusiasm for the Buckeye Bullet. On May 26, in one of his frequent open letters to newspaper editors across the United States, Rogers summarized the weekend's sporting highlights:

> Los Angeles, May 26 — The sporting pages were where the news was this Sunday morning.
>
> Dear old Babe Ruth, God bless him, stepped into three fast balls and put 'em all out of bounds. Lawson Little, a Stanford boy, for the second year in succession collected a small portion of the British debt with a golf club.
>
> A Mr. Owens, a colored lad of 21 years from Ohio State University, broke practically all the world records there is, with the possible exception of horseshoe pitching and flagpole sitting.
>
> Congress laid dormant, Hitler was refueling and Mussolini was changing records. But a man in California sued his wife for non-support.
>
> Yours,
> WILL ROGERS

The next day, Rogers, though he'd never met Jesse Owens, still liked him. In another letter to newspaper editors, he lamented three Supreme Court decisions that had seemed to undermine the New Deal—"a Republican holiday," he wrote—and concluded with these words: "So the Supreme Court just stole the spotlight from Jesse Owens."

For his part, Owens had left Ferry Field in a hurry after his remarkable hour on the track. He had no plans to spend the night celebrating. Instead, Charles Riley was waiting for him in his Model T. As they drove through the night to Cleveland, along the shores of Lake Erie, they marveled at what Owens had just achieved and considered the challenges that remained. Riley had a sense, even if Owens did not, that Owens's life had changed forever. The boy in whom he had recognized some athletic potential had turned out to be the greatest running and jumping talent the world had ever seen. But Riley knew that for a twenty-one-year-old black man in the United States in 1935, nothing was guaranteed. Over the engine noise, he praised Owens but also endeavored to keep him humble.

For a moment Owens might have thought he was hearing his father, who lived his life with his head down, always careful not to seem too proud in the presence of whites. Riley, he knew, just wanted him to be realistic. After all, there was no real money to be made in track and field. If he was going to find opportunity and security, he would have to continue to excel athletically. "The Olympics," Riley told him, his knuckles white against the wooden steering wheel, "that's what matters. Representing your country. Records will be broken, but they can never take away gold medals."

"I know, Pops," Owens said, stretching his back muscles as best he could in the confines of the old Ford. "The Olympics. I'll be ready."

"You mustn't wear yourself down before Berlin," Riley said. "You must be prepared to say no. Everyone will want you at their meets. You can't overdo it."

"But I've got plenty of time," Owens said. At twenty-one, he

still thought a year was an eternity. "And I'm only getting stronger. Look what I did today, with a bad back."

Now he was all but screaming. The car had no muffler, and
Riley had no hearing aid.

"Your injury helped you focus," Riley countered. "It helped
you."

"It helped? I could barely walk when I got out there."

As they talked, Owens and Riley could see the lights of the
tankers moving up and down the lake.

"But then the pain went away, didn't it?"

"I suppose," Owens said, wondering where Riley was leading him.

"My point, Jesse, is that you must maintain focus," Riley said.
They were crossing into Ohio now.

"I know," Owens said, "I know." But he barely apprehended
all the obstacles he would encounter. How would he maintain
focus when he had a child to support, exams to pass, and rivals
to overcome? Mostly, by doing what great athletes must always
do in similar circumstances. He would practice an icy selfishness — at least through the Olympics.

Shortly after Riley and Owens arrived in Cleveland on Sunday morning, the media descended on the Owenses' apartment.
"How does it feel to be the world's fastest human?" a reporter
asked.

"I think the praise is a little too high," Owens memorably replied. It was more than false modesty, though it was a little of
that, too. Mostly it was true ego. Owens thought the praise was
a little too high because he knew he could run faster and jump
farther. It seemed impossible to him that he might have peaked
in Ann Arbor, on a day when his back was sore and the competition far from world-class. Someday, he thought, the praise would
be just high enough.

Fairly obscure outside Cleveland before the Big Ten meet,
Owens was soon stunned to find himself celebrated everywhere.

In particular, the thriving black press elevated him to its highest pedestal, which he shared with Joe Louis, the heavyweight challenger. Both Owens and Louis were twenty-one years old, both had been born into rural poverty in Alabama, both had been part of the great northward migration to the industrial Midwest, both became stars in 1935, and eventually both would register their greatest victories in defiance of German extremism.

Even Harvard wanted Owens. The host of the prestigious Intercollegiate Amateur Athletic Association of America meet, which was to take place a week after the meet in Ann Arbor, the university invited Owens to participate. Larry Snyder said thanks, but no thanks. "After last Saturday's performances," he told reporters, "it may sound paradoxical to say Jesse is not in the best of shape, but that is the fact. He is taking heat treatments for the [back] injury, but rest will prove the big healer, and that's what he's going to get."

It had gone largely unnoticed that in addition to breaking the world records, Owens had broken another record in Ann Arbor: Larry Snyder's fifteen-year-old Ohio State record for points accumulated in a single track season. But Snyder noticed.

"Now you've done it," he said when they were back in Columbus. "You've wiped me off the books."

"Don't worry, coach," Owens said, smiling his handsome, toothy smile. "You'll see. I'll make you famous." Together they laughed, both confident that he was right.

That week, Snyder saw to it that Owens was elected team captain. As obvious as the choice might have seemed, no black man had ever before been captain of any Ohio State varsity team. Snyder didn't like it that Owens and Albritton could not live on campus — or in many places off-campus in Columbus — so he tried as best he could to make the track team, at least, a place where his black athletes would not feel the sting of segregation.

By training and inclination, Snyder was a daredevil. During World War I he had been a flight instructor, and after the war he had performed in air shows. Just riding in a primitive biplane required rare courage; pushing one of these planes to its limits was asking to die. But when it came to Owens, Snyder was cautious—not overcautious, but cautious enough. He loved speed, and so of course he loved Owens, so much that he never wanted to see his perfect machine crash.

Nevertheless, just two weeks after Owens's feats at Ferry Field, Snyder had him back on the track, and in the spotlight, in Milwaukee, where more than 10,000 people attended the Central Intercollegiate track meet. It was the largest crowd in the ten-year history of the event, and "most of [them] had come to see Owens in action," Wilfrid Smith wrote in the *Chicago Tribune*. They also hoped that they would see Owens run against the brilliant Ralph Metcalfe in the 100-yard dash, a confrontation that had been hyped in the national press as an opportunity for Owens to avenge an ancient perceived snub.

The story goes that Metcalfe, the 1932 Olympic silver medalist in the 100-meter dash, had rudely dismissed Owens a few years earlier at the national interscholastic meet in Chicago, refusing to shake his hand and turning his back on him. It seems unlikely that Metcalfe would have been so ungracious, but by the spring of 1935, as Metcalfe and Owens developed a rivalry, the story had become wire-copy gospel. The papers reported that Owens had been "almost heartbroken" and that he had made a vow to Charles Riley: "Someday I'll run that big lug right off the track." A year before the 1936 Olympics, track writers were all too eager to depict America's two best sprinters as engaged in a blood feud.

They and the rest of the crowd in Milwaukee were exceedingly disappointed when the Owens-Metcalfe showdown failed to materialize—Metcalfe pulled out to study for his law school exams at Marquette University. It is also quite possible that he was a little awed by Owens's recent heroics and had no desire to be shown up in his own town. Despite his absence, those in attendance had little reason to complain. Skipping the hurdles, Owens won the 100- and 220-yard dashes and the broad jump, once again surpassing 26 feet. As Smith put it, "There was no doubt of his superiority." Owens jumped 26 feet, 2½ inches — half a foot less than his leap in Ann Arbor, but still farther than any other human being had ever jumped. Once again the smoothness of his running style was impressive. With a typically weak start in the 220-yard dash, "he came into the stretch in fourth place. Then, almost effortlessly, he turned on the power." He ran 21.8 without breathing hard. It must have occurred to Metcalfe as he pored over his constitutional law notes that it would be easier to ace his exams than to best Owens.

That impression would be reinforced over the next two weeks, as Owens proved again and again that Ferry Field was no fluke and that the question in Berlin would be not whether he would win a gold medal but how many he would win.

Two days after the Milwaukee meet, Snyder and Owens and nineteen other members of the Ohio State track-and-field team were in Clovis, New Mexico, en route to Los Angeles, where they would face Dean Cromwell's formidable USC Trojans, featuring the sprinter Foy Draper. It was just the first stop on a prolonged swing through California that would culminate with the NCAA championships at the end of the month. For Snyder, it was a chance to show the western track powers that he had built a worthy program at Ohio State, and in particular to prove that his team could compete with Cromwell's. At the time, the fifty-five-year-old Cromwell was to college track and field what Knute Rockne had been to college football. Bow-tied and fastidious, he had built a dynasty at USC that had produced Olympic champions in 1920, 1924, 1928, and 1932. He had coached the great Charley Paddock. Snyder was eager to sic Jesse Owens on him.

In Clovis, Owens and his teammates spent a few hours stretching their legs, running and jumping on a local track, shaking twenty hours of riding the rails out of their limbs. The next day they stopped at the Grand Canyon. Meanwhile, in Los Angeles — which was still a minor-league sporting town — Owens's imminent arrival was treated as big news. In the *Los Angeles Times*, columnist Bill Henry took to calling Owens "the Cleveland Cataclysm," and each leg of the Buckeyes' cross-country journey was dutifully detailed. Finally, on June 11, Ohio State pulled into Los Angeles, "under-cover of the semi-darkness that prevails at 6:30 a.m." On the front page of that day's *Times*, in an item titled "The Negro Wonder," Harry Carr marveled at Owens's athleticism and suggested that blacks, like American Indians, were going to continue to dominate whites in sports. "Perhaps the Indians are right," he wrote. "They say that the world is coming to another Ice Age and the gods have selected certain races to survive the ordeal."

Owens had never before seen the West Coast and immediately took a liking to it. He loved the warm weather — in which it was so easy to get loose — and the sunshine. A few hours after stepping off the train and three days before the meet, he worked out at USC's Bovard Field. The *Times*'s correspondent was impressed by his affability — he called him "a good-natured colored boy" — and physique. "Looks?" he wrote, "Well, Jesse is 5 feet 10¼ inches tall and weighs 164 pounds stripped. He has probably the most perfect pair of legs that anybody ever had but he is beautifully proportioned with a fine pair of shoulders and narrow hips. There's nothing bulgy about him."

Several hundred gawkers showed up to see Owens and only Owens go through his training paces. The other athletes were "getting no attention." Black Hollywood in particular was thrilled to have Owens in its neighborhood. "The local colored colony went into a spin when they heard that Jesse was coming here," Bill Henry wrote. "They are prepared to just about turn the town over to him. And, come to think of it — why shouldn't they?"

Among those who crowded Owens for pictures and presented their hands for shaking was thirty-three-year-old Lincoln Perry — to Owens's astonishment. The first major black film star, Perry was better known to the world as Stepin Fetchit. When practice was over, Perry took Owens with him to the set of *Steam-*

boat Bill, the movie he was making for Twentieth Century Fox. Owens was mesmerized by the scene at the soundstage. When he and Perry walked onto the set, John Ford, the director, was talking to his star, Will Rogers.

"Well, look who Lincoln's dragged in," Rogers said, looking up from his script.

Perry made the introductions. "Jesse Owens, Will Rogers," he said. "And this is the boss, John Ford."

"Pleasure to meet you," Owens said to both men.

"I know you," Rogers said in his Oklahoma twang. "I've been following you."

Owens was starstruck. Stepin Fetchit was impressive enough, but Will Rogers was one of the biggest men in America. A veteran of vaudeville, a featured player in the Ziegfeld Follies, he was by 1935, at the age of fifty-five, an accomplished actor, a famed humorist, and the nation's preeminent social commentator. (He had been nominated for governor of Oklahoma, but declined to run.) To Rogers, though, Owens was the celebrity.

Recalling his brief meeting with Owens in another of his letters to newspaper editors, Rogers wrote,

> Setting here on the running board of my car, about a hundred people around, we are all trying to make a movie to make the world laugh. Step and Fetchit [sic] just come up with Jesse Owens, the Cleveland colored boy of Ohio State, who breaks world's records as easy as the rest of us break commandments. He is a very, very modest fellow. Says he will be tickled to death if he can just win these events here Saturday as he thinks these are the best boys he has met. He is entered in four events. He hold the world's record in three of 'em and is tied for the other. Funny thing, on the picture with us is Jim Thorpe, our greatest all-around athlete of all time.

It was true. Jim Thorpe was working on *Steamboat Bill* as an extra. In fact, Thorpe worked as an extra on seventeen films that were released in 1935, almost always playing a simple, noble Indian, a less offensive but no less facile stereotype than Stepin Fetchit's Uncle Toms. History does not record whether Owens and Thorpe made each other's acquaintance on June 12, 1935. It seems likely, though, that they did. Thorpe, like Rogers, was an Oklahoman, and it's hard to imagine Rogers resisting the temptation to introduce his fellow Okie to the young star whose recent exploits he had so closely monitored.

When Owens wasn't hobnobbing with Perry, he was often out partying among the black elite. Snyder had always encouraged him to relax on the nights after he competed. "It was good for him to bust loose once in a while," Snyder wrote in 1936. "He loved his Saturday nights. I used to tell him, 'Put on your glad rags and get going. This is your night.'" In Los Angeles, Owens apparently had several enjoyable nights. He met a young woman named Quincella Nickerson, the daughter of a wealthy real estate broker who was one of the pillars of the black community. Beautiful and sophisticated, Nickerson escorted him around Los Angeles, showed him the finest shops, danced with him at the swank black nightclubs. Owens seemed unperturbed by the photographers, whose snapshots were soon reproduced in several of the black papers and of course came to the attention of Ruth. In his victory-induced euphoria, he was simply oblivious of the consequences of his actions. He felt bullet-proof — and as far from Cleveland and his obligations as could be.

The day after he met Stepin Fetchit and Will Rogers — and dined at the Nickerson residence — Owens was back on the track, again training, as sportswriters sized him up and presented with scientific certitude their half-baked explanations for his greatness. "Short muscles in the calf of each leg, enabling him to get more leverage and hence greater spring, have helped to make Jesse Owens the greatest athlete in track and field history," Braven Dyer wrote in the *Los Angeles Times*, which was covering Owens's visit minute by minute. Dyer continued,

The average human being has long muscles extending almost to the heel. Jesse's muscles do not begin until halfway up his leg. The tendon is, therefore, abnormally long and the Negro star's limbs are as shapely as those of a Follies girl. In fact, there is further evidence of the feminine touch about Owens, who, however, is an athlete of exceptional strength, in addition to speed. The boy's skin is almost silky and he wears a small shoe, size 7½. His feet are neither flat nor abnormally large, as is the case with many Negroes and his heel does not jut out to any noticeable extent. One of Jesse's great grandparents was white.

The last point, oddly, was not expanded upon or ever suggested anywhere else.

Clearly Owens's performance in Ann Arbor had piqued the public's curiosity, and the men of the press box were determined to quench it. As Owens posed for pictures and jogged around the track with Foy Draper, his future Olympic teammate, Snyder was asked why he was so good.

"Nobody can answer that to the satisfaction of all concerned," Snyder said, declining to attribute Owens's talent to his race or physiology. "I know he has perfect rhythm. I know he is strong. I know he trains diligently and loves to run." Snyder paused. Then he said, "Owens came to Ohio State a great athlete. It has been my job to keep him one."

After an exhausting week of interviews and introductions, Owens was not at his best at the Coliseum on June 15. He won the four events in which he competed — handily — but did not set any personal bests. Heeding the entreaties of Braven Dyer in the *Times* — "Don't miss today's appearance of Jesse Owens at the Coliseum. The boy is the athletic marvel of the age" — more than 40,000 fans watched as Owens took the 100- and 220-yard dashes, the 220-yard low hurdles, and the broad jump. But he did not push himself too hard. "He won so easily it was hard to believe what your eyes told you," Bill Henry wrote. After Owens's final victory, in the hurdles, a swarm came down from the stands to be near him, including two black girls. One, Muriel Foley, age eleven, had him sign a soda cup. The other was shooed away by the police before she could acquire an autograph. "Well, I got to touch him anyway," she said.

After the meet, Snyder told Owens to have some fun. As usual, Owens did what he was told.

Frank Wykoff, whose 9.4 seconds for 100 yards was still the official world record — Owens's time in Ann Arbor had not yet been certified — was among the throngs at the Coliseum. The former USC star and two-time Olympian was awed by what he saw and expressed what most who saw Owens expressed: disbelief at his effortless velocity. "I never saw a man run with such ease," Wykoff said. "He doesn't appear to be running fast at all. He isn't half trying, yet he gets there first. Those strides of his are so short, and still it doesn't appear that he moves his legs fast."

Wykoff was probably one of the few people at the Coliseum hoping Owens would not break the 100-yard record, but he knew that while his record had survived the day, its end was nigh. Owens, he could plainly see, was costing himself half a second, maybe more, at the gun. "I have to get my lead in the first forty yards of the race," he said, "and then try and hold off my opponents. Owens appears too slow getting away now, but when he improves on that, look out, records."

Owens lingered in Los Angeles for several days. His next meet, the NCAA championships, at Berkeley, wouldn't begin until the following Friday. He sensed that this meet would be different from the rest. So did Snyder. For once, victory was not assured. This meet would be a challenge, because Eulace Peacock would be there.

Like Owens — and like Joe Louis — Eulace Peacock had been born in Alabama, in his case the town of Dothan, a metropolis compared to tiny Oakville. Eleven months younger than Owens,

Peacock also made the great northward trek with his family. Instead of choosing the industrial Midwest, though, the Peacocks settled in Union, New Jersey, where Eulace's father became a tar tester and Eulace did to the New Jersey state record book what Jesse did to Ohio's. Like Jesse, Eulace was a sprinter and a broad jumper. Most remarkable, as a high school student in 1933, he set a state record in the broad jump that would last forty-four years (he leaped 24 feet, 4¼ inches).

Taller and more muscular than Owens, Peacock was also more versatile. He was an excellent running back at Temple University in Philadelphia and an outstanding pentathlete. More than anything else, he relished competition. All world-class sprinters run better when they're running against the best. Peacock was an entirely different — better — runner when he was charging down the track elbow to elbow with Jesse Owens. Which is not to say that he was anything less than brilliant in Owens's absence.

On April 27 in Philadelphia, at the venerable Penn Relays, Peacock had won both the 100-meter dash and the broad jump, essentially duplicating Owens's simultaneous achievements at the equally venerable Drake Relays in Des Moines (Peacock's 10.6 in the 100 meters was judged to be roughly equal to Owens's 9.5 in the 100 yards). And on May 18, one week before the Big Ten meet at Ann Arbor, Peacock, competing in a dual meet against Villanova, ran 100 yards in 9.5 seconds, only one tenth of a second slower than Wykoff's world record.

Now, at the NCAA championships on June 22, Owens and Peacock would not have to compare their achievements telegraphically. They would face each other for the first time in the outdoor season. In the months since they had last raced against each other, Owens's reputation had grown while Peacock's had more or less plateaued, despite his wins in Philadelphia. (Compared to Owens, everyone seemed to be standing still.) Now Peacock would have the chance to prove that he too was worthy of the laurels that were being thrown at his rival, whom he had both defeated and lost to in the past.

Snyder could see that Peacock, more than Metcalfe, was the greatest threat to Owens's hopes of winning Olympic gold medals. He told Owens as much and impressed upon him the psychological importance of defeating Peacock whenever possible. "Don't let him think he can beat you," Snyder told Owens in Berkeley. "If he beats you now, he'll know he can do it again. Don't let that happen."

"Coach, nobody can beat me now," Owens said. He could be boastful with Snyder, saying things he would never allow himself to say to the writers.

In Berkeley, Owens reinforced the growing impression that he was a singular performer. He bested Peacock easily, winning the 100-yard dash, in which Peacock finished second, and the broad jump, in which Peacock finished fifth after having his best jump disallowed. Owens also won two events in which Peacock did not compete, the 220-yard dash and the 220-yard low hurdles. Perhaps most impressive, in the hurdles Owens defeated Glenn Hardin of Louisiana State. Racing in his worst event against Hardin, eventually the world-record holder and Olympic champion in the 400-meter intermediate hurdles, Owens won rather easily, despite his poor form — a photograph of the race shows him jumping several inches over a hurdle, almost as if he were high jumping — and his protestations to the contrary. "I feel fine," Owens said after the meet, "but it sure felt good when that low hurdle race was over. The competition was mighty hot in all those events today and I'm mighty lucky to have been able to win the four I was in." The Associated Press reported that Hardin "trailed Owens by a couple of yards."

As a team, Ohio State finished second, behind USC, with Owens scoring 40 of the Buckeyes' 40.2 points. Eighteen thousand fans cheered Owens's every move, which by now was invariably worthy of front-page treatment. "He beat just the pick of the country's athletic stars in every event in which he appeared," the *Chicago Tribune* reported on its front page. "From the time he won the hundred yard dash until he loped off the field after making a special effort to exceed the world's broad jump record the throng's plaudits were for Owens. No greater ovation has ever been accorded a visiting athlete."

Owens's remarkable day at Berkeley made him "the super-athlete of modern times," according to the Associated Press, and no one cared to disagree. While Snyder was exhausted — the travel and the pressure had worn him down — Owens celebrated by staying out all night dancing. "Jesse told me, 'Don't you worry about me, coach,'" Snyder told the press the day after the meet. "'You go to bed. Me, I'm going out and relax a little.'"

Owens's feet should have been aching and far from a dance floor, but since Ann Arbor he had felt nothing but elation, having tasted nothing but victory. Adrenaline was coursing through his veins, and there seemed no good reason to rest.

On June 27, just a few days after the NCAA championships, Snyder and Owens were back in Southern California, in San Diego, for a less prestigious meet, the Far Western AAU championships. Still, Peacock and Draper would be there — and most of the winners from Berkeley. Owens decided not to run the obscure 150-yard dash — in which some had hoped he would break Charley Paddock's world record — and concentrated on the 100-yard dash and the broad jump. Somehow Draper beat Owens in one of the heats for the 100-yard dash, but in the final, Owens reestablished the natural order. He finished first, just ahead of Peacock. "There were only nine inches between the two sepia speedsters," one reporter noted, "but where Peacock was run out, Owens merely extended himself sufficiently to insure no other man hitting the tape ahead of him. There wasn't the slightest semblance of effort on his part at any time." At this point, Peacock had become nothing more than a foil for Owens, a rabbit, a pacesetter, less an adversary than a means for Owens to assess his superiority.

The broad jump was even less challenging. Jumping twice, Owens reached only 24 feet, 5½ inches, which was still more than a foot farther than anyone else.

Owens's achievements were for the first time transcending sports. Suddenly he was more than just a sprinter. And he was more than just a celebrity. He was, in the parlance of his times, a credit to his race, as the *Los Angeles Times* suggested in an odd June 28 editorial titled "The Submerged Tenth":

> Jesse Owens, Negro college student, out-ran, out-hurdled, out-jumped the pick of white scholastic athletes of the country last week.
>
> Joe Louis, 21-year-old Negro, out-boxed, out-generaled and out-fought the former heavyweight white champion of the world [Primo Carnera].
>
> Paul Laurence Dunbar, Negro, charmed the world with his Mammy poems. George Garner is an outstanding tenor among the tenors of the country. The name of Booker T. Washington will live as long as education. The best fighters in the Spanish War were Negroes.
>
> [Negroes] are among the most loyal of all Americans. Most of them are quiet, unassuming, wholesome people. They have intelligence, stamina and courage.
>
> Uncle Sam can count on them.

It was the first time Owens got top billing over Booker T. Washington.

As much as he enjoyed his stardom, there were drawbacks. He had failed to realize that he was now fair game for the gossip columns, especially in the black press, which followed him closely. His prolonged flirtation with Quincella Nickerson was now a matter of public record. They had been seen together and photographed together, and it was reported that Jesse had proposed. "So important is this man," the editor of the *Cleveland Call and Post* wrote, "that his love affairs have become front page news in the daily papers."

Naturally, Ruth Solomon wasn't pleased to see the father of her daughter linked romantically to another woman — a wealthy, beautiful woman, no less — 2000 miles from Cleveland. Hurt and humiliated, she fired off a missive to Jesse that arrived just as the meet in San Diego was ending. She threatened to sue him for breach of promise — for not honoring his promise to marry her — a relatively common legal action for scorned women in the 1930s. Even after he called her to assure her that the stories were untrue, Ruth told reporters that Jesse had "one last chance to explain himself" before she would file suit. Slowly it dawned on Jesse that he had caused Ruth much pain. Then it dawned on him that his indiscretion might cause him much pain.

Their mission on the West Coast accomplished — three meets, eight events, eight victories, the elevation of Ohio State to the ranks of the national track-and-field powers — Snyder and Owens boarded a train for Lincoln, Nebraska, the site of the AAU's national championship meet. In Lincoln, Ralph Metcalfe would be waiting. Despite all his recent successes, Owens still considered Metcalfe the standard to measure himself against. His victories had meant less to him because none of them had been won head-to-head with Metcalfe. It was clear to everyone else, though, that Metcalfe now had to prove that he could compete against Owens. "Possibly Ralph Metcalfe will beat Jesse to the tape in one or both sprint events in the national championships," John Kieran wrote in the *New York Times* on June 30.

But as the country unfolded before his eyes, Owens could barely focus on Metcalfe. Instead, he could not stop thinking of Ruth and the lecture he would undoubtedly receive when they next spoke.

Amazingly, even *Time* magazine — which in those days barely acknowledged the existence of American blacks — addressed the Quincella question. "Fortnight ago [Owens] was reported engaged to one Quincella Nickerson of Los Angeles," *Time* reported. Then the magazine quoted Owens, who ungallantly downplayed the relationship. "We were at a party and Miss Nickerson asked to see my fraternity pin," Owens said. "I took it off and handed it to her, then stepped into another room for a minute. When I came back she was wearing it. Just before I left the West Coast I asked her for it. She cried a little but handed it over. I got out as fast as I could." In Owens's time, pinning a girl was essentially asking for her hand.

For once, Owens wasn't fast enough; the Nickerson controversy followed him all the way to Nebraska. He conceded that Quincella was lovely. "However," he said, "how could I have become engaged to her? Why, there was never any sentimental discussion between us. I only knew her for three days and certainly one could not grow so serious in so short a time. At least I couldn't. I'm not that sort of chap." Owens's statements also indicate that he was not, in fact, already married to Ruth, as he would sometimes later claim. If he had been, it would have been quite simple to refute the suggestion that he had proposed to Nickerson.

In Lincoln, Owens's adrenaline finally ran out. For weeks he had been feeling invincible, virtually weightless, but now, as he moved physically and chronologically closer to Cleveland, he was preoccupied with the Quincella controversy and deeply afraid that his hometown papers would report that he had had a daughter out of wedlock. Sensing that his mind and his body were exhausted, he decided to skip the hurdles and 200-meter dash and compete in only the 100-meter dash and broad jump.

It was the Fourth of July and the heat was oppressive, but still 15,000 fans filed into Memorial Stadium to see the great Owens square off against the great Ralph Metcalfe in the 100 meters. Metcalfe was still considered the world's fastest human, but Owens, of course, based on his recent performances, was considered nearly his equal. In the final, Owens crouched into his set position eleven times, and eleven times someone jumped the gun — including Owens, twice. (Under today's rules, he would have been disqualified.) His ability to concentrate had been compromised by his personal issues. The fine rhythm he had displayed in race after race for more than a month was gone. He felt uneasy. Finally, on their twelfth try, the 100-meter finalists were off. To the astonishment of the crowd, Jesse Owens could not keep up. Eulace Peacock ran the distance in 10.2 seconds — a world record if not for the strength of the tailwind — and Ralph Metcalfe, who had dreamed the night before that he would finish third, finished second. Distracted and morose, Owens settled for third.

Snyder hated to see Owens lose; he especially hated to see him lose to Peacock and Metcalfe at an important championship. "Look," he said to Owens after the race, "I know you've got a lot on your mind, but this is what matters now. You have to concentrate. You cannot let Peacock or Metcalfe think you are vulnerable. Never!"

"I'm just tired, Larry," Owens said, staring at his shoes. He said "Larry" only when he was down.

"Well, get it together," Snyder said, declining to object to what he knew was a lie. "You can't rest on your laurels. Not now. Get it together for the broad jump."

In the broad jump, Peacock jumped 26 feet, 3 inches, the best jump of his life, and Owens jumped 26 feet, 2¼ inches — a remarkable effort considering his state of mind, but still too short. Head to head against Eulace Peacock, Owens had gone 0–2. Nobody was supposed to outsprint or outjump Jesse Owens, and Peacock had done both.

"This was to have been Owens's show," Arthur Daley wrote in the *New York Times*, "and Peacock took the play right away from him. The meet was built around the Ohio State flash and not a championship did he take."

In the *Los Angeles Times,* in a note under the unfortunate heading, "Enter Mistah Owens, Exit Mistah Peacock," Braven Dyer wrote, "Peacock, like Owens, is only a sophomore and has his best years ahead of him. He is stronger physically than Jesse."

"What's wrong, Jesse?" Snyder asked after the broad jump.

"What's wrong?" Owens said, a bit defensively. "I just jumped more than twenty-six feet. Good enough for a world record. Nothing's wrong."

"Jesse," Snyder said, "I know something's bothering you. What's the situation at home? How's Ruth?"

"It's fine, coach, no problem." Owens slid from defensiveness to obstreperousness. "Everybody's spending too much time reading the papers," he said. "Ruth and me are just fine. Just fine."

Snyder was weary of these evasions. "Now, Jesse, I know you're tired," he said, "but there's no way you should be finishing third in any race, against anyone. Whatever's going on between you and Ruth is your business, but you better get it sorted out. You're not going to win any races with your mind stuck on your troubles."

"Okay, coach, I understand."

Since the day they had met, Snyder had made Owens acutely aware of how much was at stake every time he competed — nothing less than his education and his livelihood. For the Frank Wykoffs and Foy Drapers of the world — Jesse's white rivals — amateur athletics could remain just that: amateur. For Ralph Metcalfe, too, track was not everything. He was book-smart and a natural politician. Owens was neither and knew it.

Snyder's lecture in Lincoln made an impact. Owens knew that to win at the highest level, to qualify for and triumph at the Olympics, he would have to dedicate himself body and soul to the track. There could be no distractions. No more nights out on the town. No Quincellas. Berlin was just thirteen months away. Nothing else mattered. It was time to get serious.

"You've got two things ahead of you this [coming] year," Snyder said. "Randall's Island [the site of the Olympic track-and-field trials] and Berlin. And they're bigger than your Saturday nights." Of course, he was contradicting himself. He had always encouraged Owens to decompress after expending himself physically. But the situation had changed. Snyder did not fear the toll that partying would take on Owens's body. He did not fear the alcohol or the hard wood of the dance floors. Suddenly, in the wake of the Nickerson episode, he feared the girls.

At 5:40 P.M. on July 5, 1935, Owens arrived at Union Terminal in Cleveland, but Ruth wasn't there to meet him. As he hurried home to change, she was hurrying to the station. Finally they found each other.

"Well, Ruth, are we getting married tonight?" Jesse asked.

"If you got the ring," she replied.

A romantic Jesse was not—nor was he in possession of a ring. But he located one, and within a few minutes a license clerk named Frank Zizelman was summoned from his dinner table. The suddenly eager couple then called on Nelson Brewer, a probate judge, to waive the mandatory five-day waiting period between the filing of a certificate and the actual taking of vows. The Reverend M. S. Washington, Ruth's preferred minister, could not be found, but the Reverend Ernest Hall offered to perform the ceremony—as soon as choir practice was over. When the singing ended, the vows were exchanged. In addition to their friends and relatives, the *Cleveland Call and Post*, a black weekly, congratulated the couple. "Minnie Ruth Solomon won the Cupid's Sweepstakes, walking off with the prize, a yearling, one Jesse Owens," the paper reported. "Quincella of the Los Angeles Nickersons finished out of the money. A baseball scorer would be forced to credit the California lass with an assist, for she certainly helped no end in Jesse making up his mind to enter the Matrimonial Derby."

A less delicate Associated Press report appeared in the *Los Angeles Times:* "The wedding set at rest reports Miss Solomon might sue Owens for breach of promise because he was 'going around with' a wealthy Negro society girl in Los Angeles when he was there two weeks ago."

Regardless of the circumstances, Jesse and Ruth were dazzling in their wedding finery. In a wire photo, they can be seen smiling broadly, the Reverend Hall towering over them both, looking slightly disheveled.

The honeymoon was brief. Jesse was up at dawn on July 6 to catch a train to Crystal Beach, Ontario, where that afternoon he would again face off with Eulace Peacock in a 100-meter dash.

This time he surged to the lead, but with about 20 meters to the finish line, Peacock passed him, banging elbows, and outsprinted him for the win. Clearly Owens was exhausted, physically and emotionally, and was relatively easy prey for Peacock, who nevertheless was staking a serious claim to the title of world's fastest human.

The Ontario meet took place on Saturday. On Tuesday, Owens and Peacock would race again, at 100 yards, at New York University's Ohio Field in the Bronx. Proving the fickleness of the media, a small item in the *New York Times* noted that Peacock would race but did not so much as mention Owens, who just a fortnight earlier was being hailed as the greatest trackman ever. After consecutive losses to the imposing Peacock, Owens's aura of invincibility had evaporated — as had the sense that he and he alone embodied America's Olympic hopes.

Racing under a darkening sky in the summer twilight, on a track made slow by incessant rain, Peacock and Owens took their marks alongside USC's Draper, Edward O'Sullivan of the New York Curb Exchange, and George Anderson of San Francisco's Olympic Club. There were several false starts before the field broke clean, with Peacock, in lane two, zipping into the lead. At the 50-yard mark, Owens, in lane four, was at least a full stride behind. "Then," Arthur Daley wrote in the *Times*, "Owens began to move with that catlike smoothness of his that is so deceptively fast. On he came and up he came, closer and closer to the high-striding Easterner. He was at Peacock's shoulder ten yards from the red worsted, and only six inches, or less, separated the two at the tape."

Peacock had won again — his third victory over Owens in the sprint in six days. Unaccustomed to defeat, Owens struggled with the loser's protocol. He did not know when and how to congratulate Peacock, so he simply watched as the writers and the other runners came up to slap Peacock on the back. Owens liked Eulace — liked him enough, anyway — but he was not quite up to the task of feigning graciousness. This was not the lawn tennis club. This was business. Regardless of what he might say, his body language gave him away. He tried to hold his head high but couldn't. He slouched and shuffled away from Peacock and the reporters. At least he was comforted by the knowledge that he was truly exhausted. The trains, the turmoil, the wedding — all of it had sapped the strength from his legs.

The syndicated columnist Paul Gallico was in the Bronx that afternoon, braving the elements to catch a glimpse of Owens and Peacock. "It looked to me as though Owens was getting faster as the finish came closer," he wrote, "and in another step would have passed and beaten Peacock. However, by that time he had run out of cinder track and the race was over." As Peacock and Owens crossed the finish line at virtually the same time, Peacock's massive legs seemed to dwarf Owens's. Next to Eulace, Jesse was slight.

Finally, though, in the broad jump, Owens righted himself. He jumped only 23 feet, 9 inches — the track was essentially mud — but Peacock could do no better than 23 feet, 6½ inches.

When the meet was over, Owens told a radio reporter that he was fatigued and not in his best form. The reporter asked for specifics. "I just mean I'm tired," Owens said, pulling his scarlet Ohio State sweatshirt over his head. "I've been doing a lot of running and I'm tired of it. I mean, just like when you've been announcing for a long time, your voice gets tired."

His victory in the broad jump meant Owens's losing streak was over, but the sprints were what truly mattered, and on that front his abilities were being reconsidered. Even Charley Paddock, the Olympic champion who had inspired Owens as a youngster to pursue track, was suddenly a Eulace Peacock fan. "I can only see Peacock as a certain performer at the games in Berlin," Paddock told the *Los Angeles Times* after the race in the Bronx. "I'm afraid Metcalfe can't hold up another year, what with that bad leg, and I can't help feeling that Owens is pretty much burned out. If Owens were to specialize on the 220-yard dash next year," he continued, "and leave the 100 and the low hurdles alone, he still might figure. Certainly he can't continue to run in three events and broad jump besides and hope to make the grade. As for the short dash, the Ohio State boy is far too slow off the marks to do much good in the 100."

Paddock's comments were reprinted around the country, stinging Owens. More ominously, Lawson Robertson, the University of Pennsylvania track coach who would lead the U.S. track team in Berlin, said, "Peacock is the fastest and most consistent of all our sprinters. He also has a better finish than Owens."

Robertson had had many opportunities to see Peacock race in the flesh — in Philadelphia, home to both Penn and Temple — and had long ago determined that he preferred Peacock to Owens. Owens and Snyder believed, probably correctly, that Robertson was biased in favor of East Coast athletes and Paddock in favor of those from the West Coast.

The criticisms sent Owens to the phone box to dial up Snyder, who had remained in Ohio, seeking reinforcement. "So Paddock thinks I can't start and Robertson says I can't finish," he said, his voice crackling on the long-distance connection.

"You're just tired, Jesse," Snyder said, recognizing that this was a time to soothe and encourage his protégé, not to castigate him. "You just need a break."

"I know, coach, I do," Owens said, holding back tears of frustration and anxiety.

"Trust me, Jesse," Snyder said, "this is nothing. Peacock won't be able to touch you when you're rested."

But Owens couldn't help feeling unsure of himself and his ability.

In his comments to the press, he naturally emphasized his fatigue — but he also revealed a measure of self-doubt. "Of course I'm worn down," he told the estimable John Lardner. "I've been doing a lot of running this summer. I need to lay off now and take a little rest. But Eulace is a great runner — and a very good jumper. This boy has been right behind me for quite a while. It looks as though he's more than caught up now."

In an interview with the *Amsterdam News*, the newspaper of Harlem, Owens said, "It's going to take a special man to defeat Eulace Peacock. You see, I've already reached my peak. Peacock is just now reaching his. He's a real athlete. I don't know whether I can defeat him again."

If Owens was losing some of his confidence, at least Gallico was still a fan. His column the day after the Bronx meet ran under the headline "Give Owens a Rest, He'll Beat Peacock in the Dash." "If both men were at their best and fresh," Gallico wrote, "I think I should pick Owens to take Peacock in the hundred. Don't know why, particularly because Peacock has a much stronger build and seems to be more powerful all around. But Owens strikes me as more of a greyhound, and capable of a greater nervous effort over the 100-yard stretch."

Owens's "nervous" energy was a favorite topic of sportswriters. It was widely assumed that he was high-strung, like a greyhound or a thoroughbred, and that this quality helped make him great. Snyder said Owens had "a high tension nervous system." He did. But it was once said, accurately, that "even though he churned within, he remained outwardly calm."

"Owens is a form runner," wrote Jesse Abramson, the track-and-field correspondent of the *New York Herald Tribune*. "He is the picture of relaxed ease as he sprints. He never shows any apparent effort. He always appears able to do better."

No one doubted that Owens could do better than he had done after leaving California. But he would have to wait several months to prove that his great triumphs in May and June could be replicated. Most of America's best young track stars left for Europe the day after the meet at Ohio Field. Owens headed the other way, back to Cleveland and his wife and daughter.

In the space of just a few weeks he had attained international celebrity, broken several world records, reportedly proposed to one woman, married another woman, raced in the Midwest, on the West Coast, again in the Midwest, then on the East Coast, lost three consecutive races to a powerful foe, and watched as his Olympic prospects were downgraded from sure thing to long shot. Jesse Owens was tired.

4

Heel Bones and a New Start

––––––

CLEVELAND: SUMMER 1935

A S EULACE PEACOCK was spending July in France and
Luxembourg, dining on foie gras and mussels, Jesse Owens was in Cleveland, pumping gas at a Sohio service station at
the corner of East 92nd Street and Cedar Avenue. He was also
fending off accusations that he had been paid to do nothing by
the Ohio state legislature, which had been employing him as a
page. His salary was not insignificant for the mid-1930s. During the school year he had been paid three dollars a day, and
even when the school year ended, he was compensated for expenses incurred on state business. As Owens's biographer William J. Baker wrote, "The latter clause was interpreted liberally:
Jesse had his expenses paid to and from California [for the recent
meets] with a check drawn on the state treasury."

What aroused the interest of the Amateur Athletic Union —
which had the power to declare Owens a professional, thus banning him from amateur sports events such as the Olympic trials — was Owens's employment status during the early summer of 1935, when he had been setting all those world records.
Even though he had been nowhere near Columbus, he had still
received $159 as a page *ad interim*. A hearing would be held in
Cleveland on August 12 to determine his eligibility as an amateur. If he were deemed ineligible, he probably would not make
it to Berlin.

In the meantime, even after Eulace Peacock helped diminish his achievements, the public clamored for more news of Jesse Owens. In the *New York Times*, it was suggested that Haile Selassie, the beleaguered emperor of Ethiopia, which at the time was one of only three nations controlled by blacks (the others were Haiti and Liberia), was Harlem's biggest hero—next to Joe Louis and Jesse Owens. Even as Owens worked the Sohio pump, even as he prepared to defend his amateur status, the reason for his greatness was being debated. As black athletes emerged for the first time in large numbers as national figures, the public continued to be fascinated with their physiognomy. Otherwise reasonable people talked knowingly of inherent black advantages, such as supposedly long heel bones and impossibly smooth muscles. To dispel such nonsense, Dr. William Montague Cobb decided to spend some time examining Owens.

One of America's foremost black physicians and a future president of the National Association for the Advancement of Colored People, Cobb had received his undergraduate degree from Amherst in 1925, his medical degree from Howard in 1929, and his Ph.D. from Cleveland's Case Western Reserve in 1932. His specialty was anatomy. For the better part of two days, he poked at Owens's joints, X-rayed his bones, felt his muscles, and, using a set of calipers, measured the length and girth of his legs. Most impressive, Cobb thought, Owens's calves were each sixteen inches in diameter, and he was photographed stretching a piece of rope around those gargantuan muscles. Finally, Cobb, who had said he wanted to know if outstanding athletic performance was attributable to race, revealed his findings. In a word, he said no, race did not matter.

The results of Cobb's examination were reported around the country. Cobb said that "industry, training, incentive, and outstanding courage, rather than physical characteristics, are responsible for the young Negro sprinter's accomplishments." He also said that "it is a surmise that a longer heelbone, characteristic of Negroes, gives them greater leverage." In any event, he added, Jesse Owens did not have an exceptionally long heel.

The Cobb examination was a pleasant diversion from the stress of the forthcoming AAU hearing, which would take place a week later. Stories about Owens's status were on the front pages everywhere. Even Charley Paddock weighed in. Eulace Peacock might have been his choice to win at the upcoming Olympics, but in this matter he would defend Owens. A veteran of several skirmishes with the AAU himself, Paddock dismissed the charge of professionalism as "a tempest in a teapot . . . Even if Jesse is an amateur athlete," Paddock wrote in one of his occasional newspaper columns, "he still has to eat."

After meeting on August 12 to discuss the Owens matter, the local AAU chapter took nearly three weeks to announce its decision. Finally, on August 31, James E. Lee, the local AAU secretary, announced, "We failed to find that Owens was paid because of his athletic ability. We have sent an official report to the national executive committee that Owens's amateur standing is not in doubt."

With the Olympics approaching, the national media had been paying close attention to the matter. After the AAU decision, the *New York Times* declared in a headline, "Case Is Closed."

Naturally, Owens was relieved to hear that his Olympic eligibility was intact. The news made his day at the Sohio station — and Snyder's day back in Columbus.

Both Owens and Snyder were less relieved to be thrust unwillingly into a national debate about whether American athletes should boycott the Olympics in Germany. A boycott movement had been gaining strength ever since Hitler had come to power in 1933, but it did not intrude on Owens until the summer of 1935. On August 23, the *Amsterdam News* urged black athletes not to take part in the games. "Humanity demands that Hitlerism be crushed," an editorialist wrote, "and yours is the opportunity to strike a blow which may hasten the inevitable end."

Owens had no desire to be pulled into a political struggle, especially one that might cost him a trip to the Olympics. Working at the state capitol, he had had enough of politics. For the moment, he kept quiet.

Of more interest to him was the national debate about his relative greatness. Alan Gould, the AP sports editor, gave Peacock top billing in a report on American Olympic prospects one year out from the games. Gould's colleague Edward J. Neil, however, tried to dispel the notion that Owens was washed up. In fact, the headline in the *Los Angeles Times* read: "Owens Not Washed Up, Says Ohio State Coach." Actually, Larry Snyder wasn't quoted anywhere in Neil's story. Explaining Owens's losses to Peacock, Neil wrote, "It is significant that Jesse showed, very plainly, that he was putting out a lot more energy — nervous energy in the form of tension — in losing than he had in his winning races. The pictures show it. He's tightened up, laboring, as he finishes a foot or so behind Peacock. He was 'too tired to relax,' evidently. It would have been better probably if he'd quit a week or so earlier. But he can get rested up next winter." In summation, Neil couldn't resist chiding Owens's multiplying but unnamed detractors. "As for his being permanently burned out," he wrote, "that's just a lot of sand in somebody's picnic potato salad."

As always when something was going wrong, Owens went back to the basics, back to the fundamentals that had been set down for him by Charles Riley. But this time the problem lay in the fundamentals. Owens's start, the start he had always used, wasn't working. It was finally clear that he must address his greatest weakness. This was a blow to a man who had always run, and won, effortlessly. Unnerved by the knowledge that he could be caught, he enlisted Riley to help him develop a new starting position. This was not a minor development; it was, rather, a story treated as seriously as a presidential policy shift or a major stock market fluctuation. Riley, of course, was eager to help. He knew that ultimate victory was assured if Owens could improve himself at the gun.

Together, Owens and Riley spent day after day back at the track at East Tech. In the withering July heat, Riley had Owens crouch for him, then spring to his feet, a thousand times. He watched and watched and watched until the solution became clear to him.

"Listen, Jesse, this is what we'll do," he said. His old-fashioned collar was stained with sweat. "You're not going to get all the way down anymore. We must modify your set position."

With this objective in mind, Riley taught Owens to start from a semicrouch rather than the standard kneeling position in which a sprinter's fingertips touched the ground. He had Owens angle his torso at ninety degrees from his lower body, plant his left foot about 12 inches in front of his right one, hold his right arm high behind his hip, and crook his left arm in front of him. The new method, Riley told a reporter, was a combination of the standing position used by sprinters before the turn of the century and the contemporary kneeling position. "Jesse can really throw himself at the track now," Riley said. "He won't need ten, fifteen yards to dig himself out of a hole."

"Tests so far," the Associated Press reported, "show that using the conventional kneeling start Owens's first step as he comes up out of the holes is twenty-three inches. Using the new method, his first step is thirty-six inches long." The papers reported that Riley said the new start enabled Owens to cover as much ground in seven strides as he had previously covered in eight strides. But eventually Owens and Riley abandoned their unorthodox plan and worked simply on improving Owens's drive from the kneeling position.

Serenely secure in his position at Ohio State and in Owens's life, Larry Snyder was not jealous of Riley and never interfered in the older man's relationship with his protégé. As the leaves began to turn, though, it became clear to Snyder, Owens, Riley, and everyone else that the greatest threat to Owens's Olympic gold medal prospects was not his inadequate start, nor Eulace Peacock, nor Ralph Metcalfe. It wasn't the AAU's ethics committee, either. And it certainly wasn't any speedy German or Dutchman or Englishman. It was, instead, a colorful Tammany Hall judge by the name of Jeremiah T. Mahoney.

PART
II

5

The Judge and the Millionaire

NEW YORK: 1935

ODDLY ENOUGH, the most vigorous and effective pro-
ponent of an American boycott of the 1936 Olympics in
Germany was not a Jew. Instead he was a devout Irish-American Catholic known all his life for his stubborn opposition to
racial and religious discrimination. Born on Manhattan's East
Side in 1878, Jeremiah Titus Mahoney worked his way through
New York University—where he played football, baseball, and
lacrosse and high-jumped—and then NYU's School of Law. In
1923 Governor Alfred E. Smith appointed him to the state supreme court, where he served for six years before returning to
private practice. Eventually Mahoney ran for mayor of New
York, won the Democratic nomination, and lost the general election to Fiorello La Guardia by nearly 20 percentage points.

Mahoney's foil in the Olympic boycott movement, Avery
Brundage, was also a self-made man in the Horatio Alger mold.
Born in 1887 in Detroit, Brundage was forced to help support his
family when his father walked out on them. He sold newspapers
on the street. He then worked *his* way through the University of
Illinois, where he put the shot, high-jumped, and speed-walked
(in those days, race walkers were called heel-and-toers). He also
wrote for *The Scribbler,* the campus literary magazine. In a piece
titled "The Football Field As a Sifter of Men," Brundage wrote,
"No better place than a football field could be chosen to test out

a man. Here a fellow is stripped of most of the finer little things contributed by ages of civilization, and his virgin nature is exposed to the hot fire of battle. It is man against man, and there is no more thorough mode of exposing one's true self."

No one could say that Avery Brundage lacked earnestness.

By the mid-1930s, Brundage was a multimillionaire. Construction and investing had become his games and patience his virtue. "You didn't have to be a wizard to make a fortune during the Depression," he said years later. "All you had to do was buy stocks and bonds in depressed corporations for a few cents on the dollar and then wait. I was just lucky."

Even Brundage's detractors — who were legion — found him admirable in some respects. John Lardner called him "the Noblest Badger of them all." And Red Smith wrote, "He was sincere and honest and inflexible and intransigent, with an integrity equaled only by his insensitivity." Then again, Smith wrote, "Although Avery was frequently wrong-headed, he could also be arrogant and condescending."

In the 1930s, at the same time that Brundage was mastering the markets, Mahoney, who succeeded Brundage as the president of the Amateur Athletic Union, became one of the most powerful men in all of sports. In 1935, after long reflection, he came to the conclusion that American participation in Hitler's Olympics would serve only to legitimate a wholly evil regime, a regime that was discriminating against its own Jewish citizens as it chose its Olympic teams. "There is no room for discrimination on grounds of race, color, or creed in the Olympics," Mahoney said. "The A.A.U. voted in 1933 to accept an invitation to compete at Berlin in 1936, provided Germany pledged that there would be no discrimination against Jewish athletes. If that pledge is not kept, I personally do not see why we should compete."

Despite Brundage's assurances to the contrary, anyone could see that the Third Reich had no real intention of allowing Jewish athletes to compete fully on its Olympic teams. Almost since the day the Nazis had come to power, it had been clear that they planned to discriminate against Jewish athletes, despite their assurances to the contrary. Those assurances had first been offered in Vienna in June 1933, at a meeting of the International Olympic Committee. The committee had convened in part to decide whether Germany would still be allowed to host the 1936 Olympics. If the Germans refused to promise to treat Jewish athletes fairly, the committee would move the games. Initially the Germans offered merely to abide by all the laws regulating the Olympic games. "The German Olympic Committee had arrived with this promise from their government in their pockets," John MacCormac reported for the *New York Times* from Vienna. But when several American members of the IOC demanded a specific assurance that Jews would not be excluded from the German Olympic team, the German legation had to cable superiors in Berlin for instructions. Finally the Germans agreed to the broader guidelines. "What has happened is another proof of the spirit of fellowship that sport engenders," said His Excellency Dr. Theodor Lewald, the chairman of the German Olympic committee. MacCormac was duly impressed. "This development represents a complete backing down by the Hitler government," he wrote. "The straightforward character of the promise obtained from the German Government came as all the greater surprise, and the opinion was expressed that a real blow had been struck in the cause of racial freedom, at least in the realm of sport."

Of course no such blow had been struck. The Nazis, typically, simply made a promise they had no intention of keeping. Still, the IOC went to the trouble of entrusting the task of enforcing the agreed-upon regulations to Lewald and the other members of the German Olympic committee: the duke of Mecklenburg-Schwerin, Dr. Karl Ritter von Halt, Professor Carl Diem, Dr. Heinrich Sahm, and Hans von Tschammer und Osten. The Viennese reporters covering the story were skeptical. They thought, quite rightly, that "nothing but formal and empty assurances on the question of Jewish participation in the Olympics could be expected from the committee, which, it was remarked, consisted of 'diplomats rather than sportsmen.'" The Austrian press already knew how much stock to put in Nazi promises.

Just a few days after the convention in Vienna, at a Nazi party meeting in Berlin, Von Tschammer und Osten, the German minister of sport, made it clear that the Austrians were right. He told his fellow Nazis, on the record, that the pledges made in Vienna would not hinder the national agenda. "We shall see to it that both in our national life and in our relations and competitions with foreign nations only such Germans shall be allowed to represent the nation as those against whom no objection can be raised," he said. Everyone in the room knew which people were to be objected to.

Von Tschammer und Osten said virtually the same thing at another meeting, in Cologne. He wanted his fellow Nazis to know exactly where he and the German Olympic officials stood, despite Lewald's public statements. To clarify the German position for its readers, the Associated Press asked him to answer several questions. Responding to a question about a German decision to deny Jewish sports clubs "all special facilities," Von Tschammer und Osten wrote,

> It is hardly fair to expect that state support be given to purely Jewish organizations, which, being composed almost exclusively of Zionists, are even today in sharp political conflict with the government. Just as Nationalist sports organizations during the past years continued to enlist and engage in activities without any material assistance by relying purely upon themselves, so, too, no other treatment can now justly be meted out to Jewish organizations. That certainly won't create any difficulty for them, for in their circles substantial private means are available.

For three years the Germans engaged in similar rhetorical games with the international press and diplomatic corps. *No,*

they said, *we would never discriminate against the Jews. They have
every right to take part in our Olympic trials. But of course, like everyone else, Jewish athletes must be sponsored by local clubs. And of
course we cannot compel the local clubs to have them as members. These
clubs have rights, too. And they must also abide by our laws.* Which bar
Jews from non-Jewish clubs. What about Jewish clubs? *They are all
either Zionist or Communist fronts. You cannot possibly expect them
to be allowed to send athletes to our trials.* And so on.

The Germans might have been able to play this game with
the quivering western democracies and the pseudo-diplomats
of the International Olympic Committee, but not with Arturo
Toscanini. The day before the IOC meeting in Vienna, the great
conductor announced his decision to cancel his contract to direct
the Wagner festival in Bayreuth, Germany. He sent a telegram to
Winifred Wagner, the widow of Richard Wagner's son. "The lamentable events which injured my sentiments as an artist have not
yet undergone a change, notwithstanding my hopes," he wrote.
"It is therefore my duty to break the silence I have observed for
two months to inform you that, for my peace and yours, I no longer have any thought of going to Baireuth (sic)." The Germans
responded by banning Toscanini's recordings.

The same day that Toscanini made his stand, so did twenty
American Olympic champions. Together with the American
Jewish Congress, they sent a cable to the International Olympic Committee urging the IOC to stand firm against the Nazis.
Gustavus T. Kirby, the former president of the American Olympic Committee and the Amateur Athletic Union, also weighed in
against the Third Reich. "The Olympic games will not continue
to Berlin unless and until there is a change in the German attitude toward Jews," Kirby wrote. Naively, though, he assumed
that amateur athletic organizations such as those with which he
had long been affiliated might actually be able to effect change
in the new Germany. "I have a strong feeling, based on direct information that I have received from Germany, that the German
campaign against Jews will change," he went on, "particularly

if pressure is brought to bear on them." An amateur sportsman and a gentleman, Kirby apparently had no idea what kind of people he was now dealing with.

In fact, the entire relationship between the American amateur athletic community and the German authorities was warped by the Americans' fundamental misunderstanding of the Nazis. The Americans simply could not fathom that the Nazis were serious when they made speeches about cleansing German society by eliminating Jews. The (mostly) patrician American officials incorrectly came to the conclusion that Nazi rhetoric was nothing more than an instrument to control the masses. They really did not believe that the Nazis meant what they were saying.

It was also convenient for them to think this way. Their decisions were thus made much easier. For the most part, American athletes, including Jesse Owens, allowed themselves, too, to feel that the Jewish situation could not be as bleak as it sometimes seemed — on the rare occasions that they thought about the situation at all. There were many Americans, of course, who simply did not care.

But some did. In addition to Mahoney, other prominent Catholics, such as Governor Smith and Governor James Curley of Massachusetts, supported the boycott. They opposed Hitler not only because he was a racist and an anti-Semite but because his policies were also anti-Christian. In its November 8, 1935, edition, the liberal Catholic journal *The Commonweal* endorsed the boycott of an Olympics that would "set the seal of approval upon the radically anti-Christian Nazi doctrine of youth."

The boycott movement was gaining momentum. In his syndicated column, Heywood Broun wrote, "I think that one of the most useful kinds of protest that can be made against the fascist regime of Hitler lies in our staying away from the Olympic Games in Berlin."

In *his* syndicated column, Westbrook Pegler wrote, "Now that it is admitted that the German Olympics are to be a political undertaking intended to glorify the Nazi program, the American

Olympic Committee has no right to commit support to participation."

But Avery Brundage saw no evil — not in 1935, anyway. In 1933, though, immediately after Hitler became chancellor and began implementing his anti-Semitic agenda, he had sided with those who were reconsidering America's support for the 1936 Olympics, the winter games to be held in Garmisch-Partenkirchen and the summer games in Berlin. Brundage was made uneasy by the German decision to drop Dr. Daniel Prenn, a Jew who was one of the country's greatest tennis players, from its Davis Cup team. He was also troubled by reports that Dr. Lewald was to be dismissed from the Olympic committee because *he* was Jewish.

Lewald had been known to most American sports officials for decades; in 1904, at the world's fair and Olympics in St. Louis, he had thrown a party in honor of Alice Roosevelt, President Theodore Roosevelt's daughter. In Germany, he was a man of immense prestige. In 1930, on the occasion of his seventieth birthday, Dr. Lewald had been awarded the Eagle Shield, Republican Germany's highest honor, by President Paul von Hindenburg. Still, as early as April 1933, the newly installed Third Reich sought to have him removed from his post organizing the 1936 Olympics because his father, who had been baptized in 1826, was of Jewish descent. Politically, Lewald was an archconservative. In fact, he had been dismissed from the reich ministry in 1921 by his socialist superiors because they deemed him reactionary. For the Nazis, though, all that mattered was the fact that his father had been born a Jew, albeit around the time Abraham Lincoln had been born and long before the birth of the German state.

But by targeting Lewald, the Nazis overstepped, if not their authority, at least good sense. Even the usually compliant German press raised concerns. "The issue becomes all the more difficult because all experts agree the success or failure of the 1936 Olympics depends on Dr. Lewald," the *Deutsche Allgemeine Zeitung* reported. "Rome and Tokyo are making strenuous efforts to have the 1936 games transferred there. Is it worthwhile to provoke such a decision?"

The Nazis' treatment of Lewald incensed Brundage, who, ironically, was among the first Americans to suggest that the games might have to be boycotted or moved. In a letter to K. A. Miller, the managing editor of the *Jewish Times* of Baltimore, Brundage wrote, "To my mind the situation in Germany is analogous to a case that the Amateur Athletic Union encountered six or seven years ago. We had awarded the national track-and-field championships to New Orleans. Shortly before the meet we discovered that Negroes would not be permitted to compete. So we immediately transferred the event to another city." But in reality, Brundage was less concerned with German prejudice against its Jewish citizens than with the possibility that the Germans would somehow try to bar all Jews from competing — in his mind a real possibility. "If Jews were barred from American Olympic teams, I know that the AAU would voice a stern protest," Brundage, wearing his AAU hat for the moment, wrote. Then, switching hats, he continued, "And I am sure that the American Olympic Committee would do the same. Should this eventually happen I doubt that the United States would be represented in Berlin in 1936 . . . The situation in Germany may change overnight. But at any rate we are not going to permit the barring of Jews from the Olympics." This was Brundage in a nutshell. Reluctantly, he might be stirred to action. If Germany banned *American* Jews from the games, he would voice a "stern protest."

Brundage said he would see for himself how the Third Reich was treating its Jewish citizens, including, of course, its Jewish athletes, several of whom were among Germany's best, such as the high jumper Gretel Bergmann and the ice hockey star Rudi Ball. But Brundage's tour of the country in August 1934 was merely a public relations event. Before leaving for Germany, he made it clear that his sympathies were with his hosts, not with their Jewish subjects. In the *Olympic News*, he wrote, "The German committee is making every effort to provide the finest facilities and plans to reproduce the Los Angeles Olympic village. We should see in the youth at Berlin the forebears of a race of free independent thinkers accustomed to the democracy of sport; a race disdainful of sharp practice, tolerant of the rights of others and practicing the Golden Rule because it believes in it."

It was surmised, and reported, and of course true, that Brundage had already made up his mind. But the Germans were taking no chances. Hitler wined and dined the prickly construction magnate. Over the course of six days, Brundage spoke to several Jews—but only in the presence of Nazi chaperones such as Dr. Karl Ritter von Halt and Arno Breitmeyer. Not so shockingly, no one told him how bad the situation had become, and he failed to witness any overt displays of Nazi hostility to Jews. Dismissing Mahoney's concerns and changing his tune, Brundage declared that the Olympics "are an international event and must be kept free from outside interference or entanglements, racial, religious or political." He also said, "Certain Jews must understand that they cannot use these games as a weapon in their boycott against the Nazis." In other words, Brundage was saying, as he would famously say after the massacre in Munich in 1972, that the games must go on.

Even American diplomats thought that Brundage was dangerously myopic. "Should the Games not be held in Berlin," George Messersmith, the United States consul general in Berlin, wrote to his superiors in the State Department, "it would be one of the most serious blows which National Socialist prestige could suffer within an awakening Germany and one of the most effective ways which the world outside has of showing to the youth of Germany its opinion of National Socialist doctrine." It was "inconceivable," he continued, "that the American Olympic committee should continue its stand that sport in Germany is non-political, that there is no discrimination. Other nations are looking to the United States before they act, hoping for leadership; the Germans are holding back on increased economic oppression against the Jews until the games are over. America should prevent its athletes from being used by another government as a political instrument."

Still relatively sensitive to bad press and desperate to keep the Olympics, the Nazis used semantics to assuage the international community. They allowed the nomination of several Jewish athletes for the Olympic teams, but none were actually invited to tryouts. The national sanctioning bodies for each sport were to choose Germany's Olympians, but Jews were not allowed to be members of any of these athletic associations. It was a clever catch-22. To make the Olympic team, you had to be in one of the official sports clubs. To belong to one of the official sports clubs, you had to be Aryan.

Despite the obvious—and well chronicled—games the Germans were playing, Brundage went out of his way to praise their efforts to include Jews and to insult Jewish athletes. "The fact that no Jews have been named so far to compete for Germany doesn't necessarily mean that they have been discriminated against on that score," Brundage said on July 26, 1935. "In forty years of Olympic history, I doubt if the number of Jewish athletes competing from all nations totaled one percent of those in the games. In fact, I believe one-half of one percent would be a high percentage."

Most famously, Brundage absolved himself of all moral responsibility when he said that organized amateur sport "cannot, with good grace or propriety, interfere in the internal political, religious or racial affairs of any country or group."

Of course Brundage's statements were greeted with great enthusiasm in Berlin. After all, he had just given the Third Reich free rein to do as it pleased without fear of reprisal from the American Olympic Committee.

Seven weeks after Brundage's statements, Hitler made an important trip to Nuremberg, the quaint medieval city that the Nazis considered their spiritual home. It was there, on September 15, 1935, that he announced the new anti-Semitic decrees that came to be known as the Nuremberg Laws. In an instant Germany's Jews were stripped of their citizenship, deprived of protection by the laws of the land, and forbidden to marry Aryans or to employ Aryan women as servants.

At the time, the United States was far from a bastion of equality itself. Jim Crow was still very much alive and well, and not just in the South, and that made it difficult, even hypocritical, for Americans to lecture the Germans about their attitude toward Jews. But whereas race relations in the United States were for the most part getting better, Germany was clearly heading in the opposite direction. In a mere thirty-three months in power, the Nazis had turned back the clock to the Middle Ages for the country's Jews. Hitler had decided that the Olympics would serve as a spectacular showcase for his regime, so now it was all the more important that Jews be excluded from the games.

Even before Hitler's thoughts turned to the games scheduled for 1936, he had been impressed by the Olympics. The Greek Olympics, in fact, were the inspiration for the enormous party rallies that took place in Nuremberg beginning in 1927. "As I wished as many towns as possible—big, medium, and little—to participate and to become centers of German cultural life," Hitler later said, "I chose Nuremberg for our rallies, and our annual gathering there must, I think, give the city for ten days the atmosphere of the Olympic festival of ancient days." The pseudo-Greek pageantry of the modern Olympics appealed to him. He always had a weakness for pagan pomp.

On March 16, 1933, six weeks after he became chancellor, Hitler met with Lewald, already the chairman of the Olympic organizing committee, and Dr. Heinrich Sahm, the mayor of Berlin and the vice chairman of the organizing committee. According to the committee's official report, Hitler "welcomed the allotting of the Games to Berlin" and "would do everything possible to ensure their successful presentation." He also asserted that the games "would contribute substantially towards furthering understanding among the nations of the world." Of course, there was nothing Hitler more fervently desired than international understanding.

After seizing the reins of state, Hitler started transforming the fields ringing Nuremberg into a vast ceremonial plain, an expanse of stadiums, official buildings, and parade grounds that would dwarf anything that had preceded it. "This sacred site with its unique concepts of architecture and use of space," a Nazi party spokesman said in 1935, "will be the highest symbol of National Socialist life and National Socialist culture; in it the unique style of National Socialism will find its strongest expression."

So it is not surprising that Hitler was entirely disgusted by the initial plans for a fairly modest and modern Olympic stadium in Berlin. In his first six months in office, despite official statements to the contrary, he showed little enthusiasm for the games, which were still three years away. His mind was elsewhere. But it was only a matter of time before the Chancellor, who, as Albert Speer makes clear in his memoirs, loved nothing more than to immerse himself in the minutiae of architecture, inserted himself into the Olympics debate, which was soon no longer a debate. On October 5, 1933, Hitler visited the 350-acre plot of land that had been selected to become the site of the Olympic complex, accompanied by Werner March, the architect who had been chosen to create it; Interior Minister Wilhelm Frick; Von Tschammer und Osten, the new minister of sport; and Lewald. The area was known as the Grunewald, within the city limits of Berlin but eight miles west of the center. Hitler immediately nixed March's plans to rebuild the existing stadium on the site — it had been built by his father, Otto March, twenty years earlier on the grounds of a racetrack — and ordered him instead to build a new stadium that could seat 100,000 spectators. "The stadium must be built by the Reich," he said, his temper flaring. "It will be the task of the nation. If Germany is to stand host to the entire world, her preparations must be complete and magnificent."

Five days later, Otto March presented Hitler with a blueprint for the new stadium. Clearly, March was unfamiliar with his patron's aesthetic sensibilities. The stadium he envisioned was modernist, with enormous glass panels attached to the exterior, more in the style of Mies van der Rohe or Le Corbusier than Mussolini. Hitler threw a tantrum. In Speer's presence, he told Hans Pfundtner, one of Frick's deputies, to cancel the Berlin Olympics. According to Speer, Hitler said that the games "could not take place without his presence, since the Chief of State must open them. But he would never set foot inside a modern glass box like that."

"Overnight," Speer wrote, "I made a sketch showing how the steel skeleton already built could be clad in natural stone and have more massive cornices added. The glass partitions were eliminated, and Hitler was content. He saw to the financing of the increased costs; Professor March agreed to the changes, and the Olympic Games were to be held in Berlin after all — although I was never sure Hitler would actually have carried out his threat or whether it was merely a flash of pique, which he often used to get his way."

Speer's attempts to make the stadium grander were only partially successful. From the outside, it was drab, low-slung, and decidedly frumpy. On the inside, it was much more impressive and monumental. The contrast was a result of the decision to build most of the stadium underground, so that it appeared from the outside much smaller than it actually was. If one walked toward the stadium from the east, the Berlin side, it seemed almost cozy. Cozy was not Hitler's style. In fact, his taste for monumental architecture reached its apotheosis in his directive to Speer, eventually the Third Reich's court architect, to create a stadium in Nuremberg that would seat 405,000. (Speer hoped the structure would be finished by 1945, but no cornerstone was ever laid.) Hitler often said that once Germany had subdued its enemies, the Olympics' permanent home would be Nuremberg. "In 1940, the Olympic Games will take place in Tokyo," he said to Speer as they inspected a wooden model of the never-to-be-built mega-stadium, "but thereafter they will take place in Germany for all time to come, in this stadium."

It has been pointed out, by figures from Thomas Mann to Speer, that above all else Hitler was an artist — not a great artist, but a tortured, frustrated artist, whose greatest passions were opera, painting, and architecture. Not only did he design his own homes and the Reich Chancellery, he even drew the blueprint for the bunker in which he eventually took his own life. Party rallies were orchestrated as scenes from Wagner. Spectacles were the order of the day in the new Germany, and few spectacles could match the Olympics, with their rapt crowds, teams marching in unison, uniforms, flags, anthems, medals, and fatuous rhetoric. To Hitler, the Berlin games were a fascist fantasy come true — and of paramount importance in several respects.

Most significant, they would be a grand opportunity to show off the gleaming new capital he had been building. He had replaced the filth, decay, and decadence of Weimar with grand boulevards and marble palaces. Poverty had been all but eliminated. Most Berliners — at least those who had not yet been consigned to special yellow benches in far-off corners of public parks — were buoyant, floating on the tide of economic and patriotic renewal fostered by the Chancellor. The Olympics would be an important moment for Hitler and Germany, because for two weeks in February in the Alps and then for two weeks in the summer in Prussia, the eyes of the world would be trained on them. Of course, it simply wouldn't do if Germany failed to perform well at its own games. Since the revival of the Olympics in 1896 in Athens, Germany had been one of the great underachievers of the games. No German man had won a gold medal in track and field, ever. And at that time, for most people, track and field *was* the Olympics.

Strangely, in a roundabout way, Germany was the driving force behind the revival of the Olympics. Pierre de Coubertin, the French nobleman who founded the International Olympic Committee and ran it for forty years, was inspired to rekindle the games by the German victory in the Franco-Prussian War of 1870–1871. Coubertin blamed France's humiliation on the moral and physical decline of its youth. He hoped the Olympics would encourage young Frenchmen to spend less time loafing and more time running, swimming, and shooting. Ultimately, his most cherished hope was that they would be less easy prey for the Germans. By the 1930s, though, near the end of his life, Coubertin had fully embraced pacifism, like many Frenchmen who had seen the flower of French youth perish in the trenches from 1914 to 1918. The man who had created the Olympics as a hedge against German invasion now insisted that the games must be above politics. Coubertin was too blinded by his love for his creation to realize that it was the Nazis, not those agitating for boycotts, who were using the Olympics for political aims.

From the beginning, Hitler knew the games would afford him a unique opportunity to promote the thousand-year reich he envisioned. (As noted, the Third Reich was not the only fascist regime interested in hosting the Olympics. The Italians and the Japanese both made it known that they would accept the torch, as it were, if an international movement to take the games from the Nazis was successful. It should have occurred to someone that if Mussolini, Hitler, and Tojo all wanted the same thing, they must have had reasons beyond a shared love of shot-putting and dressage.)

"The Chancellor is taking an enormous interest in the Olympic Games," Sir Eric Phipps, the British ambassador to Berlin, wrote in a dispatch to the Foreign Office on November 7, 1935. "In fact he is beginning to regard political questions very much from the angle of their effect on the Games . . . The German government are simply terrified lest Jewish pressure may induce the United States Government to withdraw their team and so wreck the festival, the material and propagandist value of which, they think, can scarcely be exaggerated."

Phipps, who was sixty and a career foreign service officer, had been appointed in May 1933 to succeed Sir Horace Rumbold. In the mid-1930s, even as his government fiddled, Phipps urged Parliament to increase military expenditures to keep pace with the German rearmament campaign. He was roundly ignored. More than most observers, he knew the Nazi mentality well. To him, it seemed absurd to give Hitler and propaganda minister Joseph Goebbels — the twin masters of the promulgation of anti-Semitism, militarism, and fascism — a platform to promote their agenda. It was Hitler, after all, who increased the Olympic budget from 1.5 million to 28 million reichsmarks. It was Hitler who, in his eagerness to impress foreign visitors, diverted military funds to effect the swift completion of Berlin's new airport, well in time for the games.

Before he assumed power, Hitler had been dismissive of the Olympics because of their inclusiveness. (Any activity in which Aryans competed with Jews and blacks naturally aroused his ire.) But when he became chancellor, he changed his mind. In early March 1933, *Der Angriff*, Goebbels's newspaper, said that it was in Germany's interest to maintain its international sporting relationships. And then Hitler threw his support behind the Olympics. "I will advance the games as well as all sports interests in every manner possible," he said on March 16, 1933, the day he first met with Lewald. A month later, however, when it became clear that the regime's anti-Semitism might lead to a boycott of the games, Hitler distanced himself from the preparations. If the Jews were going to get the games moved, he was not going to allow himself to appear to have been defeated by them.

As the games approached, the Third Reich worked feverishly to make both the winter and summer games shining examples of its competency. More than anyone else, Goebbels knew precisely what was at stake. His thoughts were captured by *Der Angriff,* which wrote that the summer games would offer "a veritably historic opportunity for Germany to remove all those prejudices which have been attached to the German people not only in recent years but for decades. We are not only going to show the most beautiful sports arena, the fastest transportation and the cheapest currency; we are also going to be more charming than the Parisians, more lively than the Romans, more worldly than the Londoners, and more efficient than the New Yorkers."

As Janet Flanner put it in her "Berlin Letter" to *The New Yorker*, "For two weeks Germany profoundly wants visitors to feel at home." To make visitors feel at home, Germany had to pretend that it wasn't what it had become since 1933. Flanner reported that the brown-shirted men of the SA and the black-shirted men of the SS had been ordered to keep their uniforms in the closet as much as possible for the duration of the summer games, and that they had also been ordered not to discuss "racial problems in public, and to give a foreign lady, no matter what her profile, their seat in a tramcar." (The uniforms that would be in closets during the summer games were on full display at the winter games in Garmisch-Partenkirchen, however. During that event, foreign journalists were troubled by the omnipresence of the German military. Goebbels would not allow that to happen again.)

Like so many recent visitors to Berlin, even those disgusted by Nazism, Flanner wrote glowingly of the rejuvenation of Frederick the Great's capital. She said "the only blot" on the glittering city was "the drab" new American embassy, a token of America's financial woes in the mid-1930s. Berlin's boulevards had been beautified, its buildings repainted, its windows decorated with fuchsias. All previous efforts by Olympic host cities were amateurish in comparison. The Nazis had taken what had always been a rather clubbish, overgrown track-and-field meet and turned it into the spectacle that even now we recognize as the modern Olympics. (The 1984 games in Los Angeles were criticized for their garish, kitschy displays of patriotism. In fact, the Los Angeles organizers had nothing on Hitler and his minions.) Sounding like an official of the Berlin chamber of commerce, the eminent foreign correspondent Frederick T. Birchall wrote in the *New York Times* on July 20, "There has never been such a setting for the Olympics. Never has there been such organized landscaping, such refurbishing and polishing to show the Games at their best . . . The effect will surely be to send thousands of foreigners home with excellent opinions of the effects of dictatorship and wish that democracy might some time show itself similarly showmanlike."

At sixty-five, Birchall, an Englishman, had been the chief of the *Times*'s European bureau since 1931. With a gimlet eye, he surveyed the Nazis' Olympic preparations and aspirations — as he had observed the first three years of the Nazi reign. "All considerations of cost have been set aside," he wrote. "Every resource of German ingenuity and German organization has been brought to bear to produce equipment and setting never before attained for the Olympics. Germany has set out to show the world and she will. These games promise to be the greatest sports festival ever staged anywhere."

Of course, the games would be virtually useless to the Nazis if Jeremiah T. Mahoney succeeded in his plan to block American participation. The American athletes were the best in the world, and staging an Olympics in their absence would be akin to holding a wine-tasting competition without offerings from Burgundy and Bordeaux — a giant waste of time.

But Mahoney was blocked on both the right and the left, by the crypto-fascists like Brundage and by a large segment of the black intelligentsia. In October 1935, for instance, he ran full-speed into the double-standard argument. Speaking at Columbia University, he explained to his audience why it was a moral imperative for the United States to boycott the German Olympics. Already he was making plans for an alternative athletic festival — it would be called the People's Olympics — to be staged in (still) Republican Spain, in Barcelona.

To Ben Johnson, the black Columbia sprinter who was expected to compete for a place on the American Olympic team, Mahoney's argument was the apogee of hypocrisy. "The Negro in the South is discriminated against as much as the Jews in Germany," Johnson said shortly after Mahoney's visit to campus. "It is futile and hypocritical that Judge Mahoney should attempt to clean up conditions in Germany before cleaning up similar conditions in America."

Johnson — who, unlike the other sprinter named Ben Johnson, never tested positive for steroids — had a pretty good point. For fifty years blacks had been banned from the major leagues. Much more recently, the National Football League had drawn the color line. (After Ray Kemp was released by Pittsburgh in 1933, no blacks played in the league until 1946. Incidentally, Kemp became the track coach at Tennessee State, where his charges included Ralph Boston, who eventually broke Jesse Owens's world and Olympic broad-jumping records.)

Additionally, in 1935 no one could have foreseen the horrors awaiting European Jewry. Many German Jews themselves were still optimistic that things would get better, not worse. Johnson could not be blamed for equating the plight of America's blacks with the plight of Germany's Jews. At that time Hitler had probably murdered more Nazis than Jews, most notably during the so-called Night of the Long Knives, in 1934, when he had purged the party of some of his oldest comrades, including Ernst Roehm, the brutal chief of the SA, or Brown Shirts. Kristallnacht, the infamous state-sponsored pogrom, was still three years away.

But Johnson did not speak for all black Americans. For instance, in August 1935, the *Amsterdam News*, which was published only a few blocks from Columbia University, had already urged Olympic hopefuls to take a stand against fascism by staying home. Not surprisingly, though, most athletes — including the Jewish sprinters Marty Glickman and Sam Stoller — were against the boycott. The black athletes rationalized their decision by pointing to domestic prejudice and, like the Jewish athletes, by suggesting that winning in Berlin would embarrass the Third Reich and repudiate its claims of racial superiority.

In early October, Mahoney announced the formation of the Committee on Fair Play in Sports. Its mission was stated clearly on its letterhead: "No American Participation in the Olympic Games under Nazi Auspices." Committee members included Heywood Broun, Governor Curley, Reinhold Niebuhr, Adam Clayton Powell, Jr., and Norman Thomas.

Meanwhile, the anti-boycott position was laid out nicely in a letter to the sports editor of the *New York Times* by a reader named George S. Schuyler. "With the possible exception of the Scandinavian countries," Schuyler wrote,

> where could the Olympic Games be held where liberty is not stamped into the mud and millions are not ruthlessly persecuted and exploited? Right now it probably would be impossible for the much-lauded Ralph Metcalfe, Jesse Owens, Eulace Peacock, Cornelius Johnson, Al Threadgill, Archie Williams, Ed Gordon, Jimmy LuValle or any of the other sepia-tinted stars to get a decent meal or a room in any public hotel or café below the Mason-Dixon line (and but few above it), to say nothing of actually competing in a track meet with the Southerners. Why does the Fair Play Committee remain silent about this condition? Must we move our jim-crow areas to Germany before these gentlemen will break their silence?

Schuyler's points — and Ben Johnson's — were quite fair but in the final analysis irrelevant to the AAU's decision. Brundage and his lieutenants were no more interested in the civil rights of black Americans than in the civil rights of German Jews. They were mostly concerned with the perpetuation of their own power. And there was little power to be wielded by an Olympic committee that could not send a team to the Olympics.

6

"We Are with You, Adolf"

A S BRIGADIER GENERAL Charles Sherrill of the American Olympic Committee disembarked from the *Normandie* in New York Harbor on October 21, 1935, after seven weeks in the fatherland, he was met by William Chamberlain, an aide to Mahoney on the Fair Play Committee. In full view of the assembled reporters, Chamberlain handed Sherrill a copy of an angry letter Mahoney had written to Lewald. Sherrill accepted it grudgingly and tucked it into his jacket pocket without reading it. He had been warned that Mahoney might attempt such a stunt.

A former U.S. ambassador to Turkey, Sherrill had been in Germany on September 15, when the Nuremberg Laws were announced. If he had been paying attention, he would have seen the Hitlerites at their worst to that point. Still, his views remained unchanged, and like his friend Brundage, he declared that the games must go on. "I went to Germany for the purpose of getting at least one Jew on the German [winter] Olympic team," Sherrill said,

and I feel that my job is finished. As to obstacles placed in the way of Jewish athletes or any others in trying to reach Olympic ability, I would have no more business discussing that in Germany than if the Germans attempted to discuss the Negro situation in the American south or the treatment of the Japanese in California. I am surprised at the extent to which the movement to keep America out of the Olympic Games has gone in this country. I am sorry that what I have done has not pleased all of my Jewish friends, many of them the most prominent Jews in New York. But I shall go right on being pro-Jewish, and for that reason I have a warning for American Jewry.

Remarkably, Sherrill had the audacity to tell America's Jews that he was motivated, at least in part, by a concern for their safety. He said that if American Jews continued to agitate for a boycott, there would surely be a backlash. "There is grave danger in the Olympic agitation," he said.

Consider the effect of several hundred thousand youngsters training for this contest throughout the United States, if the boycott movement gets so far that they suddenly are confronted with the fact that somebody is trying to defeat their ambition to get to Berlin and compete in the Olympic Games. We are almost certain to have a wave of anti-Semitism among those who never before gave it a thought and who may consider that about 5,000,000 Jews in this country are using the athletes representing 120,000,000 Americans to work out something to help the German Jews.

Sherrill pointedly refused to acknowledge that Jeremiah T. Mahoney was no more Jewish than he, even as the Fair Play Committee insisted that its goal was to pressure the Third Reich on several fronts, not just its institutionalized anti-Semitism. "The issue is not Germany against its Jewry," Chamberlain, Mahoney's deputy, said, "but fair play. It has been denied not only to Jewish athletes in Germany but also to Catholic and Protestant sports clubs that do not accept Nazi doctrines of conscience."

That night, in Yonkers, just north of New York City, Mahoney dismissed Sherrill's rationalizations and maintained his attack on the Nazis. Meanwhile, as Mahoney was holding forth, Sonja Branting, a prominent Swedish judge, was speaking on Manhattan's West Side, at the West End Synagogue. Decidedly gentile, she praised Mahoney and insisted that there would be more protests if the Olympic games were held in Berlin.

The following day the controversy continued to boil. Attacks were launched and countered. Frederick W. Rubien, the secretary of the American Olympic Committee, even out-Sherrilled Sherrill. "Germans are not discriminating against Jews in their Olympic trials," he declared, ridiculously. "The Jews are eliminated because they are not good enough as athletes. Why, there are not a dozen Jews in the world of Olympic caliber—and not one in our winter sports that I know of."

That in itself was an astonishing statement. Rubien must have known that Irving Jaffee, a Jewish speed skater, had won two Olympic gold medals at the Winter Olympics in Lake Placid in 1932. No other American had ever won two gold medals at a Winter Olympics. In fact, twenty-three Jewish athletes had won Olympic medals—not merely competed—at the winter and summer games in 1932. Additionally, in 1935 Jews were well represented in all major sports in the United States. That year Hank Greenberg of the Detroit Tigers was the Most Valuable Player in the American League, Max Baer defended the heavyweight championship wearing a Star of David on his trunks, and Barney Ross won the world welterweight championship. In track and field, it was thought that young Sam Stoller might one day duplicate the feat of another great Jewish sprinter, Harold Abrahams of Great Britain, who won the 100 meters at the Paris Olympics in 1924.

In short, Rubien, a sixty-four-year-old New York City tax official, was talking like a Nazi. He too had gone to Germany, in August, and upon his return had said that there was "absolutely no discrimination" against Jews or anyone else competing for an Olympic spot. "These stories about discrimination against German-Jewish and Catholic athletes are not based on fact," he said.

Rubien also dismissed as "absolutely absurd" reports that anti-Semitic signs had been put up in the area around Garmisch-Partenkirchen, where the winter games were to be held. There were, in fact, anti-Semitic signs all over Germany, including, most famously, one that William L. Shirer, the foreign correspondent who eventually wrote the definitive English-language history of the Third Reich, saw at a sharp bend in the road near Ludwigshafen, which read:

DRIVE CAREFULLY!
SHARP CURVE!
JEWS 75 KILOMETERS PER HOUR!

In fact, as the AAU vote neared, the case against participation was being made on all fronts. As one of three American members of the International Olympic Committee — along with Brigadier General Sherrill and Colonel William May Garland — Ernest Lee Jahncke wielded considerable influence. Both Count Henri de Baillet-Latour, the Belgian head of the IOC, and Dr. Lewald of the German organizing committee asked Jahncke to use "his good offices in behalf of American participation." A former assistant secretary of the navy of German extraction, Jahncke balked. On November 26, in New York, he announced that he could not in good conscience support American participation. He made public the impassioned letters he had written to De Baillet-Latour and Lewald.

To the Belgian, he wrote:

The fact is that Jewish athletes, as a group, have been denied adequate opportunity for training and competition. Indeed, the Nazi sports authorities have themselves admitted that to be so. And The Associated Press, an impartial news service, has reported: 'In only a few German cities may Jews use public athletic fields. To build and maintain their own grounds is almost impossible because of the cost. Consequently, many Jewish sportsmen have been forced to play in the country fields and pastures where no facilities are available for many contests such as track events . . .' You quote an argument used by propagandists for holding the games in Nazi Germany — the fact that Negroes are excluded from many private clubs in America, as if what some Americans do in their own private social relations, however unfortunate it may be, were at all comparable to the treatment of the German Jews by the Nazi government and party. There is still time to arrange for holding the games elsewhere than in Germany. Let me beseech you to seize your opportunity to take your rightful place in the Olympics alongside of de Coubertin instead of Hitler.

But De Baillet-Latour chose Hitler. "The boycott campaign does not emanate from national Olympic committees and is not approved by any of our colleagues," he wrote in a letter to the American Olympic Committee. "It is political, based upon groundless statements, whose falseness was easy for me to unmask."

To combat such intransigence, Judge Jeremiah T. Mahoney was fond of quoting from a pamphlet published in 1934 by a Nazi party functionary named Bruno Malitz. In the pamphlet, titled "Sport in the National Socialist Ideology," Malitz wrote, "It is logical to expel the Jew from German sports activities. There is no such thing as a Jewish German. Moreover, we do not wish to have Negroes traveling in Germany and meeting our fine sportsmen in competition." Malitz added that "Frenchmen, Belgians, Polaks and Jews once ran on German tracks and swam in German pools. But it cannot be said that international relations between Germany and its enemies were bettered. The true spirit of German sports was destroyed. Only traitors say otherwise."

Malitz's diatribe was also cited by Alfred J. Lill, a member of the American Olympic Committee, on the evening of December 3 at an impressive boycott rally organized by the Fair Play Committee and attended by 2500 people at the Mecca Temple on West 55th Street in Manhattan (a site sacred to Shriners, not Muslims). Lill — who later defended Charles Lindbergh and assumed an isolationist position himself—was a featured speaker, as were Mayor La Guardia and Governor George H. Earle of Pennsylvania. "Have no doubt of it," Earle said, "the Nazis will take advantage of the occasion to sell their new philosophy to everyone who attends [the Olympics]."

Earle knew the Nazis well. As Franklin Roosevelt's first ambassador to Austria, he had seen firsthand the frightening power of Hitlerism and knew the Third Reich well enough to know that none of its pronouncements could be trusted. "If you want your children to be taught that might is right," he bellowed from the lectern, "that woman is a lower animal than man, that free press, free speech, and religious freedom are false ideals, that peace is weakness, that liberty as we have learned to love it in America is a myth—if you want these doctrines inculcated in the youth of America, then send your boys and girls to Germany!"

Eight other governors expressed their support of the boycott in letters read at the rally. So did seven U.S. senators. By voice vote, a resolution was passed urging the Amateur Athletic Union to boycott Berlin. The resolution charged the Nazis with "the regimentation of all sport activity, with seizing control of the preparations for the Olympic Games, and with the duplication in the realm of sport of those basic Nazi policies of regimentation and oppression which have aroused the condemnation of the civilized world."

When Jeremiah T. Mahoney spoke, he promised that he would not compromise on the issue and that at the upcoming convention where the vote would take place, he would "ask for no quarter and give none." He was engulfed in applause.

Then it was La Guardia's turn to speak. Rising from his chair —he was seated next to Irving Jaffee, the great Olympic speed-skating champion—the mayor approached the lectern. Short and squat, he cut an entirely different figure from the athletic Mahoney. He was also the son of a Jewish mother. "I came here," La Guardia said, clearing his throat, "because I want to enlist and take an active part in opposing American participation in the games if they are held in Berlin." He paused. Applause filled the silence.

> It is most important that the games are scheduled to be held under the auspices of the Nazi government—the temporary government of the German people. One of the finest things for world peace has been the Olympic Games held every four years in a different country. Athletic contests imply good sportsmanship and fair play, two qualities which are unknown to the Hitler regime. It is evident that these games are to be exploited by the present regime. Our athletes, I hope, will refuse to lend respectability to Hitler and his followers. The American people, it would seem, are not in favor of sending their athletes to the meeting in Berlin.

Rather less convincingly, La Guardia further said that "if we could ascertain the viewpoints of the athletes themselves, they would be overwhelmingly against participation."

La Guardia underestimated the single-mindedness and selfishness required of world-class athletes. But his mischaracterization of the athletes' viewpoint was perhaps the only false note struck all night. The assertion that the Nazis would use the games to legitimize their regime was entirely correct. Brundage might even have agreed. It is likely that he hoped the games would, indeed, legitimize the Third Reich, for which he had developed an undeniable fondness.

Walter White, the secretary of the NAACP, also spoke at the rally. More than ever, White was convinced that American blacks could not in good conscience compete in Berlin. He expressed his sentiments to the audience that night and in a telegram to Mahoney:

> Will you convey to the Amateur Athletic Union the very sincere request of the National Association for the Advancement of Colored People that it vote decisively against American participation in the 1936 Olympics games if they are held in

Germany? This is asked in no spirit of racial selfishness nor in any holier-than-thou attitude.

The United States has much to answer for in the matter of racial discrimination, especially against Negro athletes in the South. Instead, we ask the AAU to vote against participation on the ground that Germany has violated her pledges against racial discrimination, and for American athletes to participate would be to negate every principle upon which the Olympic games are based.

Refusal to participate will, we believe, do untold good in helping Germany and the world to realize that racial bigotry must be opposed in its every manifestation. To participate would be to place approval upon the German Government's deplorable persecution of racial and religious groups and would stultify the Amateur Athletic Union and all athletes who participate.

The rally at the Mecca Temple was a success in terms of the turnout and the passion of the oratory. But what impact it would have on the AAU's delegates was unknown. Meanwhile, the nation's best athletes had decided to do what it was clear they would always do. Despite the pressure from the NAACP, they announced that they wanted to run. The morning after the rally, papers around the country reported that most of America's most prominent black Olympic hopefuls had sent a letter to Avery Brundage supporting American Olympic participation. While self-interest was the overriding principle, their statement also reflected their dissatisfaction with the situation of blacks in the United States. The letter was signed by Jesse Owens, Ralph Metcalfe, Eulace Peacock, Ben Johnson, and Cornelius Johnson — together, five men who might have defeated any track team ever assembled. Metcalfe spoke for all of them when he said, "No political situation should alter plans for the coming games in Berlin."

As Brundage had said the previous day, "These colored athletes do not fear in Germany the racial discrimination they encounter in our own southern states." Brundage had made a habit of dissembling on the issue of Germany's Olympics, but in this case he was telling the truth.

On December 6, just three days before the AAU was to decide whether it would support American participation in the German Olympics, the German steamship *Bremen* docked in New York. Among those disembarking was a dark-haired, thickly built six-footer wrapped in a double-breasted herringbone overcoat. He was Max Schmeling, the thirty-year-old former heavyweight champion of the world. Schmeling had returned to the United States from his homeland to secure his next opponent — either Jim Braddock, the reigning heavyweight champion, or Joe Louis, the undefeated heavyweight prodigy — and to carry a message to Avery Brundage.

After checking into the Hotel Commodore, adjacent to Grand Central Station on 42nd Street, Schmeling received Brundage in his room. Brundage asked for — and received — Schmeling's assurance that black and Jewish athletes would be welcome at the games in Berlin. Schmeling also delivered the letter he was carrying from the German Olympic committee, more or less repeating its promise that it was practicing fairness in all respects. "In retrospect, it was incredibly naive of me to guarantee things that were completely beyond my control," Schmeling later wrote in his memoirs.

On December 8, Schmeling took a ride to Pompton Lakes, New Jersey, where for the first time he met Joe Louis. Schmeling's old friend Paul Gallico was among the writers covering the Louis camp in the days leading up to his fight with Paulino Uzcudun on December 13. Sitting a few feet from the ring where Louis spent the better part of an hour beating up his sparring partners, Gallico and Schmeling watched the young fighter closely. The crowd inside the small gym reflected the diversity of the fight game — whites and blacks mingling, trading barbs, exchanging handshakes. Unlike Jack Johnson, the only black heavyweight champion to that point, Joe Louis excited most white boxing fans. He was their hope, too. Surveying this scene with Gallico, Schmeling must have been reminded of the situation in Germany, which had seriously deteriorated — from the Jewish viewpoint, anyway — with the passage a few weeks earlier of the Nuremberg Laws.

As they chatted about Braddock and Louis and the state of the sport, Gallico turned to Schemling and asked a serious question. "What would old Adolf have to say if you lost to a black man?"

"I don't think he would care much." Schmeling's English was good, but his accent was thick, as heavy as the marzipan cake he so enjoyed.

"Well, would he approve of a fight against Louis?"

Schmeling paused. "The Führer," he said, his tone just shy of sarcasm, "has other things to worry about."

Moving on, Gallico said, "Well, what about the Olympics? You know we're sending over a bunch of Negro boys, and some people here think they might be in danger in Berlin."

"Don't be silly." Schmeling chuckled, the way a villainous German officer might in a Hollywood movie. At moments like these, it was easy to see why he was called the Black Uhlan. "All Germans know how good the American Negroes are," he said. "We know they will beat our German runners. It's not important that they win — as long as our Germans make a good showing."

Meanwhile, as Gallico and Schmeling went back and forth about how America's blacks might be received at the Olympics, the boycott question was being decided across the Hudson at the Hotel Commodore. Finally, after more than two years of debate, the Amateur Athletic Union delegates were deciding whether to sanction American participation in Hitler's Olympics. Technically, the AAU could not veto American participation; that power belonged solely to the American Olympic Committee. But if the AAU voted to boycott the games, the AOC would find itself in an almost untenable position. Both politically and financially, the AOC would be hard-pressed to send teams to the winter games in Garmisch-Partenkirchen and the summer games in Berlin. As the *New York Times* noted on its front page on December 7, "The principal effect of the resolution, as analyzed by committee members on both sides of the question, is intended to be moral."

There would be practical consequences as well. For instance, the AAU would not organize and conduct Olympic trials, which had always been its primary Olympics-related function. It was not the AOC but the AAU, as the master of the trials, which certified American athletes for the Olympics. If passed, the resolution would also strongly encourage, if not actually compel, those AAU members who sat on the American Olympic Committee to work toward a boycott.

For his part, Brundage was defiant. "If necessary," he said, "we will form our own organization and send a fully representative team of certified amateurs. Our athletes must not be denied the chance to carry the Stars and Stripes to victory abroad just because of treason for political reasons in some quarters at home."

The meeting was more or less a fiasco. The resolution "calling upon" American athletes not to participate in the games was tabled by a vote of 61.55 to 55.45 (some districts divided their votes and some emeritus members had partial votes), but the vote was quickly followed by an ugly parliamentary quarrel. Pro-boycott forces insisted that they had been cheated out of the two hours they said they had been allotted to speak in favor of their cause. "It was double-crossing by men who claim to be sports," Charles L. Ornstein harrumphed. "We will enter into no further agreements [with the opposition]. In that spirit I will offer an amendment to their resolution and on that we will stand and fight."

Their resolution, which had passed, called for the United States team to go to Germany. Ornstein's amendment called for the opposite, in effect an effort to negate the resolution it was amending. Naturally Brundage protested. But Mahoney, the chairman of the convention, said that Ornstein was in order. At one point during the increasingly heated debate, Mahoney addressed the floor. "I am glad the motion to prevent discussion by tabling and compromise has been frustrated," he said. "We will now have a discussion of a moral principle more important than anything which has ever come before a sports governing body." Pausing for dramatic effect, he looked out across the room. "The Nazi government," he said, "wants more than American participation in a sporting contest. It wants to bring the American dollar into the very weakened Nazi treasury. And it wants to picture Hitler with Uncle Sam standing behind him and saying, 'We are with you, Adolf!'"

This was more than a rhetorical flourish. However hypocritical the AAU leadership might have been, however laggard when it came to racial integration in the United States, Mahoney spoke the truth. After years of debate, if the United States was to send its teams to the Olympics, its action would be viewed universally as nothing less than a validation of the Third Reich, which had just stripped its Jewish citizens of their most basic rights.

Initially Avery Brundage had based his position on the principle of the separation of politics and sport. But by December 1935 he had immersed himself in international politics. More than even Father Charles Coughlin, the Canadian-born priest whose fascist leanings became most pronounced only after 1936, Brundage had become the preeminent American apologist for Nazi Germany.

After their bitter and divisive afternoon, the AAU's delegates put on their tuxedoes and sat down to have dinner together. As gentlemen, they had agreed that the Olympics would not be discussed at their gala. But to the enduring resentment of the Mahoney forces, Brigadier General Charles Sherrill stood up from his seat in the Hotel Commodore's ballroom and brandished the November 21 edition of the *Times* of London. Linking the American boycott effort with international communism, Sherrill read aloud a *Times* dispatch from Riga, Latvia: "One of the immediate tasks (of the International of Communist Youth in Russia) is to defeat the plans for holding the next Olympic games in Germany." He also read a letter from an official of the Archdiocese of New York, clearing him, as far as it was concerned, of the charge of anti-Catholicism, which had been made against him by *The Commonweal.* Then Sherrill sat back down and finished his meal.

The next day the delegates were back at it. For five hours pro-boycott and anti-boycott forces waged a pitched rhetorical battle, speaker after speaker taking the floor to exchange rebuttals. The focus of the debate was another resolution, technically a proposal by Judge Aron Steuer to send a committee to Germany to investigate conditions there, in actuality the last resort of the boycott forces to continue their fight. Mahoney and Brundage allowed their fellow delegates to do most of the talking. In addition to Sherrill, Gustavus T. Kirby, the former AAU president; John T. McGovern, the president of the Intercollegiate Amateur Athletic Association of America; and Fred L. Steers, the AAU's third vice president, spoke against the boycott. "What you are trying here," Steers said, "is not a case of sporting discrimination, but a moral judgment on Germany as a whole, which we have no right to impose on our athletes." Then Steers revealed that he and some colleagues had conducted a poll of all the American athletes who had finished first, second, or third at the most recent AAU championships and the most recent NCAA outdoor championships, as well as those who held world records in track and field. "Out of one hundred and forty replies," Steers said, "we received only one against participation. That was from Herman Neugass of Tulane University." (Neugass, a Jew, had finished third, behind Owens, in the 220-yard dash at the NCAA outdoor championships in June. He had probably had some lengthy talks about the proposed boycott with Ernest Lee Jahncke, who was the number-two man at Tulane and an old friend of Theodor Lewald's.)

All the momentum had swung to the anti-boycott forces, and the applause for Steers seemed to confirm Mahoney's fears that he would not carry the day. Brundage had outmaneuvered Mahoney by timing his supporters' final speeches so as not to co-incide with the luncheon adjournment. It was left to Jack Rafferty, the union's first vice president, to rally the pro-boycott forces, although it seemed it was already too late — the delegates were agitating for food. "A famous spell-binder," according to the *New York Times*, Rafferty started by saying he had come 2000 miles from Texas; then he lambasted his colleagues. "We have become the greatest group of hypocrites on earth," he said. He was now straining to be heard above the rattle of delegates rising from their chairs to go off to lunch. "We say we are trying to inculcate in youth the principles of fair play, and we give an example in our own councils of throttling free and open discussion in violation of an earlier agreement."

Then, to a mostly empty room, delegate Louis di Benedetto rose to speak. He said that participation would be a "betrayal of honor and sportsmanship by any Catholic, Protestant, or Jew who loves his God."

Then came the vote. If the resolution failed to pass, the boycott movement would be dead. The delegates' vote was close, 58.25 to 55.75, but Brundage won. Barring some unforeseen development at home or abroad, the United States would participate in the Olympics in Garmisch-Partenkirchen and Berlin.

Jeremiah T. Mahoney shook his head in defeat and disgust. He still could not understand why his longtime colleagues — men of honor, he thought — would elect to legitimize the most evil government in history. A few minutes after the vote, he announced that he would not seek reelection as president. "I bow to the will of the majority," he said. "But I could not in good conscience carry it out. When conditions change in Germany, the evidence will change my views. Under no consideration, therefore, can I accept nomination for any office in the Amateur Athletic Union." Mahoney then also resigned from the American Olympic Committee, which he said "was afraid to face the facts that the holding of the 1936 Olympic Games in Nazi Germany is a travesty of the Olympics ideals of sportsmanship."

With Mahoney out, Brundage was drafted and elected president of the AAU unanimously. Soon thereafter, for his boycott stance, Jahncke was kicked off the International Olympic Committee and replaced by Brundage.

As Mahoney stubbornly and implausibly vowed to continue the fight, word reached Jesse Owens that the boycott movement had been defeated. Now he knew that he would go to Berlin. There was, of course, still the possibility that he might be injured — or that somehow he would be upset at the Olympic trials. But that possibility seemed remote. Owens never thought about injuries, and it was almost impossible to conceive a scenario in which he would fail to qualify for at least three Olympic events. Oddly, Owens and Brundage, who would become enduring symbols of the good and bad of the Olympics movement, respectively, had won a joint victory. If not for Brundage's pigheadedness, cunning, Germanophilia, anti-Semitism, and deep-rooted bigotry, Jesse Owens would never have become an Olympian.

7

A Blessing in Disguise

IN THE DAYS leading up to the AAU vote, Jesse Owens, the highest-profile American Olympic hopeful, found himself in a difficult position. He knew that if the United States did not participate in the Olympics, he might miss his chance at fame and fortune. And the politics of the boycott meant little to him. The indignities suffered by German Jews seemed almost mild compared to those he and his family had endured. But when he was asked his thoughts about the proposed boycott in a radio interview in November, he had taken a moral, unselfish position. "If there is discrimination against minorities in Germany," he said, "then we must withdraw from the Olympics."

When Larry Snyder heard what Owens had said, he reacted protectively. Immediately he thought that Owens had lost his mind. He told him that the boycott would be pointless; that Germany was just the host of the games, not its governing body; and, most important, that it was too late to change the Olympic venue, and if the Olympics were held in Germany, Jesse Owens was going to Germany.

In Columbus, Snyder was criticized for challenging Owens's thinking, but he refused to be cowed. "Jesse Owens is sitting on top of the world today," he said. "If he continues to participate in this activity, he will be a forgotten man." Pointing out that Owens had not been invited to the upcoming Sugar Bowl track meet in New Orleans, he went on to say, "Why should we oppose Germany for doing something that we do right here at home?"

When Owens and Snyder sat down to talk about the situation, Owens said that he had been running all his life to escape the American variety of Hitlerism.

"All I want is the chance to run," he said one night in Snyder's office in Columbus.

"I know," Snyder said. "The boycott won't happen."

But the uncertainty of the situation made Owens anxious. He knew that success at the games might change his life for the better. He knew that a boycott would mean four more years of training without pay, all the while getting older and probably slower. When Snyder had dressed down a group of students who had wanted Owens to go on the record supporting the boycott, Owens had been grateful.

"I see no reason to get into a controversy about the Olympics," Snyder told reporters on November 10. "The games have been awarded to Germany, all preparations have been made, and now some people want to have America withdraw just because some of the German policies are not approved by them."

Snyder was a progressive. Unlike Avery Brundage of the AOC and Dean Cromwell of USC, he was far from sympathetic to the Nazi cause. But his first loyalty was to Jesse Owens. He thought that if Owens got the chance to compete, he would win every event he entered. He knew, too, that then Owens would never have to look back. Of course, it is also crucial to remember that Snyder's opinion was not informed by the gift of foresight. Like the AOC, he did not know, as we now know, that there would be a holocaust, that Hitler and his regime would eventually kill millions, that the Germans would attack Poland, France, and the Soviet Union. If he had known, he would have felt differently about the boycott. But in 1935 it was still possible to assume that European Jewry was not on the precipice of extinction, just as it was possible to believe that Hitler was not quite a madman. Everyone knew that Hitler disliked the Jews, but few imagined that he would attempt to exterminate them.

In the end, Snyder convinced Owens to support participation in the Olympics, and Owens joined the other elite black athletes in signing a letter to the AAU delegates. But that was not the end of it; some black newspapers and leaders immediately pleaded with him to reconsider his position. In his hometown, for instance, the *Call and Post* urged Owens to reverse course, calling the Nazis "the world's outstanding criminal gang."

For his part, Walter White felt that it was his personal duty to convince Owens — not Metcalfe or Peacock or Johnson, who had also signed the letter — that his initial, pro-boycott stance had been correct and that he had erred on the side of evil by flip-flopping. He sent him a telegram, which read, in part:

December 4, 1935

My dear Mr. Owens:

Will you permit me to say that it was with deep regret that I read in the New York press today a statement attributed to you saying that you would participate in the 1936 Olympic games even if they are held in Germany under the Hitler regime. I trust that you will not think me unduly officious in expressing the hope that this report is erroneous.

I fully realize how great a sacrifice it will be for you to give up the trip to Europe and to forgo the acclaim, which your athletic prowess will unquestionably bring you. I realize equally well how hypocritical it is for certain Americans to point the finger of scorn at any other country for racial or any other kind of bigotry.

On the other hand, it is my first conviction that the issue of participation in the 1936 Olympics, if held in Germany under the present regime, transcends all other issues. Participation by American athletes, and especially those of our own race, which has suffered more than any other from American race hatred, would, I firmly believe, do irreparable harm. If the Hitlers and the Mussolinis of the world are successful it is inevitable that dictatorships based upon prejudice will spread throughout the world, as indeed they are now spreading. Defeat of dictators before they become too deeply entrenched would, on the other hand, deter nations, which through fear or unworthy emotions are tending towards dictatorships. I hope that you will not take offense at my writing you thus frankly with the hope that you will take the high stand that we should rise above personal benefit and help strike a blow at intolerance. I am sure that your stand will be applauded by many people in all parts of the world, as your participation under the present situation in Germany would alienate many high-minded people who are awakening to the dangers of intolerance wherever it raises its head.

> Very sincerely
> Walter White
> Secretary
> NAACP

When the telegram reached Owens in Columbus, he read it slowly and carefully. Here he was, a twenty-two-year-old sprinter in the middle of one of the most complex controversies of the time. This was not an issue he was equipped to handle. He had none of Ben Johnson's rhetorical talents — not yet. He had focused for so long simply on running and jumping. First Charles Riley and then Larry Snyder had told him that as long as he kept practicing, he would become the greatest track star the world had ever known. No one had foreseen complications such as those he was now confronting. No one had prepared him to be a symbol, or a trailblazer, or a statesman. Yet he was being asked, all but begged, by perhaps the most respected black man in the United States to take a stand that would necessarily prevent him from achieving all that he had worked so hard to achieve. To Owens, this was the height of arrogance. Who was Walter White to ask him, even to suggest to him, not to pursue Olympic glory in Berlin?

All these thoughts were foremost in his mind as he walked to

Larry Snyder's office, telegram in hand. He wanted to show it to
Snyder, to hear once and for all that he was making the right decision. But then suddenly he stopped in his tracks. He realized
that he did not need Snyder to confirm what he already knew.
As far as he was concerned, he was going to Berlin. In this debate, and this debate only, he would side with Brundage, who
belonged to a whites-only club in Chicago, not with White, who
had dedicated his life to racial progress. In seventy-two hours
the AAU would decide once and for all whether Owens and the
others would be allowed to compete in Berlin. But Owens already knew where he stood, so he turned around and walked
back home. He would see Snyder later. At that moment, there
was nothing for them to discuss.

When the news arrived from New York that the boycott measure
had been defeated, Owens was relieved. The uncomfortable controversy was over. In all likelihood, he would have the opportunity to compete in Berlin. At that point, his spirits needed lifting,
since they had been dampened the day before by the announcement that Lawson Little, a golfer, had won the Sullivan Award,
the Amateur Athletic Union's highest honor. Little had won both
the British and American amateur championships, two years in
a row, a unique achievement, but not on a par with Owens's accomplishments at Ann Arbor alone. Adding insult to injury, the
names of all ten finalists for the award were announced, and
while Eulace Peacock was one of them, Owens, absurdly, was
not. Without publicly stating its reason for denying Owens a
place among the finalists, the AAU nevertheless made it known
that the controversy over his summer job at the Ohio statehouse
was at the root of its decision; it did not matter to the AAU that
Owens had paid back the money at issue, $159. Apparently the
episode meant Owens could not have been the American amateur athlete who had "contributed the most to the cause of
sportsmanship," as the award criteria demanded.

Owens's ego was only somewhat soothed ten days after the

Sullivan snub, when the Associated Press released the results of its nationwide poll to determine the outstanding athlete of 1935. Again he did not win. In fact, he did not even place (Little finished second). Joe Louis, who had become a top contender and the biggest draw in boxing by knocking out former champions Primo Carnera and Max Baer, took the top spot. But Owens did finish third, just ahead of Jay Berwanger, the University of Chicago running back who had just won the first Heisman Trophy (at that time called the Downtown Athletic Club trophy — it had not yet been named for John Heisman), and Mickey Cochrane, the catcher-manager of the world-champion Detroit Tigers.

But Owens's mood was soon darkened again. On December 28, when he was stripped of his athletic eligibility by Ohio State for failing his fall psychology course.

Breaking the news to Snyder was the hardest part. Over the phone. But to Owens's surprise, Snyder was encouraging, not disappointed or angry. "This is only a small setback," he said, trying to be heard over the sound of Bing Crosby on the radio. Then, brightening, "You know, this might be all for the best."

"How's that?" Now Owens was both disappointed and confused.

"We can take this time to really work together, to focus on technique, without worrying about results. Your body can use the rest, too."

"But I don't want to let down the team. I'm the captain."

"Jesse, don't worry about that. You'll be back. This will all work out for us."

When he met with reporters the next day, Owens struck just the right tone of humility, confidence, and determination. "I am disappointed," he told them. "After all, the school has done so much, and when it comes time to pay dividends and you can't, it kind of hurts . . . I haven't had much time to study, you know. I work two hours a day at the legislature and four hours a day at the gasoline station." Then, defiantly, he made a vow. "Above all things," he said, pointing his finger, "I'll pass those fifteen hours of study next quarter."

His pride was wounded. It bothered him that the whole world had been made aware of his academic failure. He had flunked the course's final examination — five essays and seventy-five true-or-false questions — and would miss all the winter meets, with the possible exception of the Millrose Games in New York, which he could enter as an independent competitor.

In February, as Owens's exile stretched into its second month, Snyder, as was his wont, tried to accentuate the positive. "We regret having Owens out of our indoor meets, but looking at it from a national standpoint, his ineligibility is a blessing in disguise," he said to a group of reporters. He was in sweats, directing an indoor practice. He spoke slowly and clearly, to have his point register. "Jesse is the greatest track star in the country, and his late start is certain to give America several points in the Olympics."

8

Jew Kills Nazi

IN FEBRUARY 1936, in the charming twin Alpine villages known collectively as Garmisch-Partenkirchen, the world got a sense of how the Nazis would manage the games of the Eleventh Olympiad in Berlin in August. The Fourth Olympic Winter Games were a splendid celebration of snow and ice, dominated, as always, by the ladies' figure-skating competition, which Sonja Henie won for the third consecutive time. Norway's answer to Sweden's Greta Garbo, Henie charmed the Führer and everyone else. Her elegant figure-eights were the perfect counterpoint to the goose-stepping in the snowy streets. Compared to the previous games, held in Lake Placid, New York, in 1932, in an atmosphere uncharged by politics and militarism, the games in Garmisch-Partenkirchen were a fascist show. Regular army soldiers were constantly parading through town. Black-shirted members of the SS were the Olympic security force. They were everywhere, demanding credentials, blocking access, intimidating journalists and spectators.

William L. Shirer was among the journalists who lamented the situation. "Too many S.S. troops and military about," he wrote in his diary, "not only for me but especially for Westbrook Pegler! But the scenery of the Bavarian Alps, particularly at sunrise and sunset, superb, the mountain air, exhilarating, the rosy-cheeked girls in their skiing outfits generally attractive, the games exciting, especially the bone-breaking ski-jumping, the bob races (also bone-breaking and sometimes actually 'death-defying'), the hockey matches and Sonja Henie."

It was to the great good fortune of Pegler's readers that he decided to cover the winter games. Casting his jaundiced eye on the Third Reich for the first time, he described things most of his fellow newspapermen ignored. In particular, he was fascinated by Hitler, not the spellbinding Führer he had read about but the plain, dumpy human presence he found. "So far as his appearance goes," Pegler wrote from Garmisch on February 20,

> it is not hard to imagine him back at his old trade, standing on a scaffold painting the side of a house, but we know the man who now stands on a reviewing platform receiving the patriotic and personal adulation of more than a hundred thousand Germans is a tremendous power, and it makes your flesh squirm on your bones to reflect that a nod or shake of his head can start a war that would kill millions of men or prevent a war and spare their lives. All this power resides in the person of a man who weighs about 200 pounds and wears a belted down leather overcoat and a brown military cap and smiles much more than you would expect him to. He has rather heavy eyelids and from the set of his mouth under his small mustache it would seem that he had his share of trouble with his teeth.

Hitler was not in Garmisch for the duration of the games. On February 13, for instance, he was in Schwerin to eulogize Wilhelm Gustloff, the forty-one-year-old physicist who was his hand-picked Nazi party organizer in Switzerland, who had been shot to death nine days earlier. Gustloff's killer, David Frankfurter, probably was not thinking about the winter Olympics when he pumped five bullets into the Nazi in Gustloff's study in Davos, Switzerland. True, the games were only two days and 100 miles away. And yes, Davos was a winter sports resort, too, a stronghold of the snow-and-ice set, where the Olympics must have been a popular topic in the lodges and cafés. But Frankfurter was a single-minded and serious young man. A twenty-six-year-old Croatian-born Jewish medical student, he had traveled by train from Bern, the Swiss capital, to Davos a week before the killing to allow himself some time to learn Gustloff's routine. After killing Gustloff, Frankfurter, who later said his plan had been to commit suicide immediately, instead calmly put down his gun, walked to a police station, and turned himself in. He killed Gustloff because he thought it would be too difficult to assassinate Hitler. "I did it because he was a Nazi agent who poisoned the atmosphere," Frankfurter said at the time of his arrest. "The bullets should have hit Hitler."

The German government and press responded swiftly to the killing. "Nazi Germany, from the bier of its shot party group leader," the *Völkischer Beobachter*, Hitler's newspaper, proclaimed, "renews its oath to proceed without compromise in preserving the German people for eternity from the slavery of international Jewry." The paper also reported that "the Gustloff murder by a Jew throws a glaring light on the rotten activity of Jews in the whole world. Germany lost hundreds of men through Jews and Frankfurter's act justifies the Nazi separation of Jews from Germany."

Under the unfamiliar headline "Jew Kills Nazi," *Time* magazine described Frankfurter as "a nervous, hollow-eyed young Jew" on February 17, 1936, in an issue whose cover photograph is still often reproduced. The photo is an action portrait of a handsome if not classically beautiful brunette in short shorts struggling, oddly, *up* a ski slope. The caption reads: "Hitler's Leni Riefenstahl."

Frankfurter picked an excellent time and place to kill a Nazi functionary. The maximum sentence in Switzerland for political murders was fifteen years, and despite vehement calls from Berlin for the death penalty for Frankfurter, the Swiss did not budge. Also, official Olympic goodwill prevented the Nazis from launching a full-scale retaliatory assault against either Switzerland or its own Jews. "Still thinking of the Olympic Games," *Time* observed, "Germany dared take no obvious reprisals, but bitter little Propaganda Minister Goebbels promptly ordered that all Jewish theatrical meetings, concerts, lectures, etc. in Germany be abolished 'until further notice.'"

At Gustloff's funeral, Hitler again blamed Jews for the hundreds of Germans who had been killed and the thousands who had been wounded in political battles since 1918. The invisible hand destroying the Teutonic tribe belonged, he said, to "our Jewish enemy, driven by hatred, who tried to subjugate the German people and make them slaves and who is responsible for all the disasters which befell Germany [since 1918]." Continuing, the Führer said, "You, Wilhelm Gustloff, have not fallen in vain. Little did your assassin dream that through his deed millions would be awakened to true German life."

Pegler, from his perch in Garmisch, took note of the absurdity of the Führer's assertion that Germany's postwar troubles were the fault of its relatively minuscule Jewish population. "It seemed a fine acknowledgement of the superior mentality of the Jews," he wrote, "and a rather melancholy estimate of the patriotic character and brains of the members of the race of supermen."

But neither Pegler nor any of the other Olympic visitors had the vaguest idea that at the same time Chancellor Hitler was playing host to the winter games, he was also making final preparations for his boldest and riskiest gambit yet. The same man who nonchalantly signed autographs at the hockey game between the Americans, the eventual bronze medalists, and the Canadians, the eventual silver medalists, had already decided that shortly after the games his forces would reoccupy the Rhineland, violating the terms of the treaties of Versailles and Locarno and all but daring the western powers to invade Germany.

At the hockey game, it occurred to Pegler, contrary to David Frankfurter's calculations, that it would take no great conspiracy to eliminate the Führer. Watching Hitler at the ice hockey rink, mixing freely with fans, unattended by security, Pegler wrote, "There was a temptation to speculate on the chance that an assassin would have had inside the arena. Hitler sat in full view, without uniformed guards, and it seemed that almost anyone on his side of the place was free to walk up and thrust an autograph book in his hand." Or, as Pegler was suggesting, a knife in his gut.

Despite the arrogance of the SS and the military displays, the Nazis were for the most part on their best behavior during the games. The anti-Semitic diatribes were temporarily toned down—though only the most willfully ignorant visitors took the Nazis' cordiality as anything other than a cynical ploy to court public favor. Everyone knew—even the American journalists who arrived only days before the opening ceremony—that there were now signs throughout Germany saying JUDEN UN-ERWUNSCHT, "Jews Not Wanted Here." But they had been taken down—"purged," as Pegler wrote, "to use a familiar word."

Meanwhile, the American team, which had endured so much uncertainty owing to the boycott debate, might as well have stayed home. After winning twelve medals in Lake Placid in 1932, including six gold medals, Team USA won a grand total of four medals—one gold, three bronze—under Hitler's gaze. The Norwegians, Germans, Swedes, Finns, and Austrians all won more; in other words, to the Chancellor's delight, the games were a testament to the virtues of Nordic manhood (and, thanks to Henie, womanhood). Only the two-man bobsled team of Ivan Brown and Alan Washbond won gold for the United States.

In a typically blunt dispatch titled "Send Over Joe Louis," Paul Gallico wrote, "Your correspondent caught a severe cold yesterday after spending five frigid hours on the mountain watching the bob run. He is striking today so he has time to review the situation and think things over. He has reached the conclusion that it would have been better if the Jews had won out and let the boys stay at home. Nothing has been done by our boys to add to American prestige, although Avery Brundage, head of the American committee, looks beautiful in a Tyrolier hat and his badges."

As the games in Garmisch-Partenkirchen were winding down, Jesse Owens was waiting out the winter 4000 miles away in Columbus. Because of his ineligibility, he had not been able to participate in any of Ohio State's indoor meets. Still, he and Larry Snyder remained constantly at each other's side. In their limited spare time, coach and pupil worked on technique and fitness. Snyder's prediction that the downtime would help Owens at the Olympic trials and the Olympics had become gospel in the U.S. track community. Writers from New York to Los Angeles pointed out that after his strenuous schedule in the winter and spring of 1935, Owens had faded in the summer heat. That would not happen in 1936, they wrote. He would be fresh at Randall's Island and in Berlin. It was a damn good thing, they suggested, that he had failed that psychology course.

Meanwhile, Owens's absence was a decidedly bad thing for the Buckeyes. At the Big Ten's indoor championships in Chicago on March 15, Ohio State finished sixth. Its only winner was Dave Albritton in the high jump.

On Friday, March 20, however, the midwinter term ended and Owens, who had reversed course academically, was eligible again. He returned to the track just two days later, at the Butler Relays in Indianapolis. Together, he and Albritton and several other teammates were driving through Indiana on their way to the meet when they stopped at a roadside restaurant. Knowing they were unwelcome, the black Buckeyes waited outside for their white teammates to bring them food. This apparently wasn't good enough for the proprietor, who stormed outside, screaming that he would "feed no niggers." Typically, Albritton wanted to fight, and typically, Owens calmed him down.

Unfazed by his second run-in with Hoosier racists in the

span of just a few weeks — there had been another ugly incident involving him and Albritton in Richmond, Indiana — Owens proved at the Butler Relays that three months of inactivity
had not taken a significant toll. Shaking off some rust, he won
all three events he entered. In the 60-yard low hurdles and the
broad jump he was not quite himself, but in the 60-yard dash he
was spectacular. Running on a slow track, he covered the distance in 6.2 seconds, only one-tenth of a second off the world record he shared with Ralph Metcalfe. "Jesse Owens has served
notice to the track world that it takes more than three months'
layoff to dull the fine edge of his abilities," the *Washington Post*
reported.

"See, Jesse," Snyder said, slapping Owens on his sweaty back,
"when you win those gold medals at the Olympics, you should
send one to that psychology professor."

"He can have one," Owens said, smiling broadly for the first
time since December. "One."

9

A Friend and a Foe Felled

A WEEK PASSED before Owens was on the track again. This time he was back home, in Cleveland, for an exhibition at the unusual distance of 50 yards. This time he would not be facing the best that the Midwest could offer. This time he would be racing the one man who had his number.

As Owens warmed up and stretched, he could not keep his eyes off Eulace Peacock. His nemesis appeared stronger than ever. For Peacock there had been no layoff, no chance for his muscles to go slack or his focus to drift. As always, he looked like he would run over—not just past—the rest of the field.

On this night, only four men would run the 50. They would use primitive starting blocks—an advantage for Owens, whose start was better than it had been but still his only weakness. As he walked to the starting line, Owens heard footsteps behind him.

"These aren't the trials," Larry Snyder said. "Remember that. I know you want to beat this guy, but don't let it affect your form—and don't go too hard. You're just getting into shape."

"I know, coach," Owens said. For all his naiveté in some matters, Owens was a careful tactician. He knew that if he ran too hard he might injure himself. With the Olympics only four months away, that would be the worst of all possible outcomes.

He wanted to show Peacock he could defeat him — but as Snyder said, this was not necessarily the time.

A few moments later, Owens crouched into his starting position. Peacock was inches away. The two rivals had exchanged pleasantries earlier. But now there was only the race.

At the gun Owens broke cleanly, but he could sense that something was wrong — not with him but with Peacock, who had tripped coming out of the blocks. Peacock scrambled to his feet, but the race was lost. He did not attempt to finish. He was still wiping the cinders from his shorts as Owens broke the tape. The crowd of 6000 was understandably disappointed. In a moment of true chivalry, Owens offered Peacock a do-over.

"Come on, Eulace," he said, "that one didn't count. Let's do it again."

"That's kind, Jesse," Peacock said, slightly stunned by his great rival's magnanimity. "Okay."

The announcement that there would be another race was greeted with a standing ovation. Just a few minutes later, all four starters were back at the blocks and the gun was fired. Still slightly winded from his previous effort, Owens faded late, and Peacock won by no more than a foot, in 5.5 seconds.

"Jesse," Peacock said to Owens in the locker room, "that was a wonderful gesture."

"Come on, Eulace," Owens responded, tearing his sweatshirt off, "it wouldn't have been right to have sent all those folks home unhappy."

"Still, I appreciate it," Peacock said, his newest trophy at his feet. In fact, Owens and Peacock had developed a true fondness for each other — despite the rivalry that might have made it difficult for them to wish each other well.

When their discussion turned to their plans for the upcoming weeks, Peacock said, "I want to run where you run. There's no point in us not running against each other. We can only make each other better for the trials."

"But I was going to go to Drake," Owens said, referring to the meet that would take place in Des Moines the weekend of April 25 and 26. "You're going to be in Philadelphia."

"Well, then, you should be in Philadelphia too." Peacock was grinning. After all, he had just defeated Owens again, his fourth consecutive victory against the man some of the papers called "the Brown Blizzard."

To the bitter disappointment of the organizers of the Drake Relays — where Owens was the reigning sprint and broad-jump champion — Owens decided to take Peacock's advice. Instead of traveling to Iowa at the end of April, he would go to Pennsylvania. Overall, the competition would be stronger, and it was time, he and Snyder agreed, to put an end to Peacock's winning streak. Pressure had also been applied by Lawson Robertson, the University of Pennsylvania track coach, who would serve as the head coach of the American Olympic team. He had made Snyder aware of his desire to see Owens in person. Of course, Owens's presence would also greatly benefit the Penn Relays, which took place on Robertson's home field and which he oversaw. In short, Robertson had much to gain if Owens agreed to run in Philadelphia.

Between the exhibition in Cleveland and the Penn Relays, Owens and Snyder spent a week together in West Virginia. They were there to visit four colleges and ten high schools in recognition of National Negro Health Week. Actually, that's why Owens was there; Snyder was there because Owens was there. In Charleston, in Huntington, in Bluefield, and in Fairmont, Snyder woke Owens up early and put him through his paces on local tracks. Owens could not afford to look out of shape at the Penn relays — not with Lawson Robertson glaring at him, not after having lost four consecutive races to Eulace Peacock.

"You know, Jesse," Snyder said one morning in Charleston as they jogged together around a high school track, "if you keep listening to me, you might be pretty good someday."

"If I keep listening to you," Owens said, "I might keep losing to Eulace."

The next day, in Bluefield, they were at another track, at another high school, and now Snyder's tone was more serious. "Jesse, I don't think anything can stop you — not even Eulace. But you've got to promise me that you'll tell me when you need a rest. This is the most important time of your life. You cannot overdo it. Got me?"

"Larry," Owens said, "I'm good. I feel really good. We will be there."

They both knew where "there" was.

During this time, Snyder and Owens grew closer than they had ever been. Traveling alone together, working toward the dual goals of success at the trials and then in Berlin, coach and athlete came to understand each other. The coming weeks and months would obviously be the most important of their lives to date. They could not have known, or even suspected, that there would be no Olympics in 1940 and 1944, but they did suspect that this would be Owens's only chance. There would be no way for him to support his family while maintaining his amateur eligibility. In the classroom, he was no Ralph Metcalfe. He would not be able to hide away at law school for three years. For Owens, the Olympics would mean either glory or a future pumping gas at the Sohio station. No one knew this better than Snyder, and the reality of the situation fueled his determination to make Owens an Olympic champion. The unfairness gnawed at him. The bigots who belittled him for recruiting and training Owens and Albritton would not have the last laugh. He would not allow it to happen.

By the time the Penn Relays rolled around, Owens was rounding into form. He was much stronger and sharper than he had been in Indianapolis and Cleveland. He was more focused, too. The Olympic trials were only ten weeks away. There was no time now for horseplay, as there had been before the Big Ten meet a year earlier, nor for dancing, as there had been in Los Angeles. The spring of 1936 was the time when Owens's natural gifts were brought fully to flower by a ceaseless regimen of training (but not too much training) and sleep. He slept ten hours a night, or more. He ate prodigiously — Snyder fed him red meat whenever possible — and he drank not at all.

None of this, however, mattered quite so much as what happened in Philadelphia. At Franklin Field on the campus of the University of Pennsylvania, the greatest obstacle to Owens's command performance was overcome.

Eulace Peacock had suffered a heel injury the previous summer at a meet in Milan. It had nagged at him for months but seemed to pose no serious threat to his Olympic hopes. But on April 24 he was called upon to run the anchor leg for Temple in the 440-yard relay. By the time he was handed the baton, Texas and its anchor, Harvey Wallender, had a significant, if not insurmountable, lead. Sprinting furiously to catch up, running harder than he should have, Peacock pulled up lame 15 yards from the tape, clutching at his right hamstring, which had all but snapped. "A look of pain and surprise spread over his features," Arthur Daley wrote in the *New York Times*. Peacock was able to stagger across the finish line, clutching his right thigh, but at that moment he knew he would never see Berlin. He realized that his life's work had been wasted.

Like many other erstwhile rivals and teammates, Owens — who had run a sparkling 220-yard leg in the medley relay to assure victory for Ohio State — tried to console Peacock, offering unfounded hopes that he might be able to recover in time for the Olympic trials. Peacock accepted the good wishes graciously, but inside he was anguished over his reversal of fortune.

For Owens, Peacock's implosion was an unexpected and wholly unwanted boon. Owens was not the kind of man to rejoice in the misfortune of a respected and feared rival. Still, now only Ralph Metcalfe was standing in his way. Frank Wykoff, Sam Stoller, and Foy Draper were simply not legitimate contenders for the sprint titles. They might fight among themselves for the chance at a bronze medal, but they would not win. Sprinting isn't like tennis or golf, individual sports in which on any given day the tenth-best or twentieth-best player in the world might defeat the best. A sprinter who runs the 100-meter dash in 10.5 seconds doesn't suddenly run the same distance in 9.9 seconds. Peacock had proved time and again that he was capable of consistently besting Owens. No one else had—not in years, anyway.

The day after Peacock went down, Owens won the 100-meter dash in 10.5 seconds, breaking Peacock's meet record but not straining himself too greatly. Sam Stoller matched him stride for stride until the final 20 meters, when Owens took off. Owens won the broad jump, too, but cleared only 23 feet, ⅝ inch. There was a good reason for his lackluster performances. Having witnessed Peacock's injury, he had decided that he would not risk everything by overextending himself at a meet that, for all its prestige, was essentially meaningless. He was heeding Snyder's advice not to overdo it. But even at half-speed, he was too good for a field that lacked Peacock, who, it was announced, would be laid up for four weeks and was now considered unlikely to qualify for the Olympic team.

In fact, just three weeks after the Penn Relays, Peacock was back in competition—but not running or jumping. In a dual meet against New York University, he displayed his strength by throwing the javelin. Without ever changing out of his street clothes, he threw it well, more than 163 feet, good for third place. He probably should have been resting.

As Peacock moonlighted in the field events, Owens picked up his pace on the track. Snyder watched gleefully as his captain recaptured his form of the previous spring. In three different meets in the first three weeks of May, Owens tied his world record in the 100-yard dash, at 9.4 seconds; ran the same distance in 9.3 seconds, with a strong wind at his back; jumped 24 feet, 10¾ inches to win a broad-jump competition; and easily won all his 220-yard races. On the blustery afternoon in Madison, Wisconsin, where he ran the 9.3, Snyder could not help noticing that everything the runner was doing seemed . . . well, perfect. His start was better than ever. His posture. His breathing. His pacing. It was almost as if, Snyder thought to himself, there was nothing for *him* to do but get out of the way.

"Jesse, you see what they're saying," Snyder said a few days later as together they prepared for the Big Ten championship meet, which was about to take place in Columbus. "They're saying you can't be as good this year as you were last year."

"Come on, coach," Owens replied, "why would they say that?" After nearly a year of doubt and turmoil, his swagger had returned. Standing on the Ohio State track, hands on hips, smiling broadly, he was the picture of confidence. Not since leaving California the previous summer had he felt so fast. But there were doubters.

"I don't know, Jesse, that's what some of the writers are saying," Snyder said. "They're saying nobody can be that good twice." He was only teasing — but he was right. That was what the writers were saying.

Francis J. Powers, the *Los Angeles Times*'s special track correspondent, had a lengthy pseudoscientific discussion about Owens with Leo Waner, a physician and once prominent runner at the University of Kansas.

"When Owens was smashing records on the Michigan track," Waner said, referring to the miracle of Ferry Field, "he was in perfect balance. His endocrine, psychic, nervous, and muscular systems all were in perfect tune, and such a blend happens but rarely." In summation, Waner said he doubted that the athlete would ever duplicate his performance at Michigan. "Owens may be in record-breaking form in one or even two events on any given day, but I doubt if he can be sufficiently perfect for four."

For his part, Powers theorized that Owens was not quite as good as he had been a year earlier, because in 1936 he was so far injury-free. "A few hurts seem necessary for him to do his best running and jumping," he wrote.

Frank Hill, who as Northwestern's head coach had been routinely victimized by Owens, also subscribed to the injury theory. He said that Snyder "should give Owens a good belt with a club before each event, just to get the boy in the proper mental frame . . . If Owens ever is hurt badly enough," Hill said, "he'll jump twenty-seven feet."

Maybe Powers and Hill had a point. Without an injury to sharpen his focus, Owens did not set any world records at the 1936 Big Ten meet, in the 220-yard dash. He did, however, manage to win the events he had won in 1935, but without breaking any records. The finals were in Columbus, on May 23, and more than 14,000 spectators came out to see him take part in the 100- and 220-yard dashes, the 220-yard low hurdles, and the broad jump. They had no complaints. Nor did Snyder. Owens's victory in the hurdles was typically ugly — and inspiring. Tripping over the first hurdle, he was in last place by the fifth; then, running the rest of the race nearly at full speed, he edged Michigan's Bob Osgood to win.

Owens had run the hurdles for his team and for Snyder, for whom the Big Ten championship was of critical importance. Of course he had no intention of running the low hurdles at the Olympic trials; it was not an Olympic event. Even as he won his four events, the Buckeyes could not hold off Indiana, which won the Big Ten title largely — and literally — on the strength of its weight throwers.

As Owens was racking up wins in Columbus, Peacock announced that he would soon be ready to run again. Every day for the next several weeks his trainer worked on his injured hamstring, hoping somehow to have him ready to compete at the regional Olympic trials at Harvard. First he raced at an AAU meet in New Jersey — but could not finish. Then, at Harvard, he struggled again, in both the 100-meter dash and the broad jump. Still, Peacock would not give up. Having been granted a special exemption to attempt again to qualify for the national Olympic trials on July 11 and 12 in New York, he went to Princeton on July 4 to participate in another regional qualifier.

Jesse Owens was also in Princeton, as was Ralph Metcalfe. It was awful to watch strong, swift Eulace Peacock hobble around the stadium with his right leg heavily taped. Owens didn't know quite what to say to him, other than "Good luck." Peacock, meanwhile, had no more luck at Princeton than he had had any-where else recently. He barely cleared 22 feet in the broad jump and withdrew from the 100-meter dash. The AAU decided, how-ever, that it would nevertheless allow him to compete in both events at the trials at Randall's Island.

If it was now clear that Peacock was a broken man, it was equally clear that Owens was again a force of nature. He won the broad jump handily, and the 100-meter dash, slightly ahead of Ralph Metcalfe.

After Princeton, Owens went home to Cleveland for a few days, to catch up with the wife and daughter he had been too busy to see. He would be leaving for New York soon and, barring catastrophe at the trials, would then be on his way to Germany for the most consequential three weeks of his life. Since there were plans for him to tour Europe with the rest of the American track team following the games, Ruth and Gloria would not see him for almost two months. Neither would the Cleveland press corps, which, for all its Owens enthusiasm, could not afford to be in Berlin, not at the height of the Depression. The Ohio trip therefore became a whirlwind of interviews and appointments. Reporters asked questions, friends offered their best wishes, and some businessmen made promises of employment that they had no real intention of honoring.

Then there was Jesse's family. Emma, his mother, was mel-ancholic but proud as she sent her son off on what would be his defining journey. "Now don't go dancing with any of them Ger-man girls," she cautioned.

Ruth was also sad to see Jesse go, but she knew exactly what was at stake: everything. "Jesse, just do what you always do," she said to him in their bedroom on their final night together. "You know you're better than anyone else. Just stay healthy and everything will work out."

In the wake of all his recent triumphs, Jesse was brimming with confidence. "Baby," he said, squeezing Ruth's hand, "from here on out, everything for us is going to change."

10

Olympic Trials

A s it happened, the Olympic track-and-field trials were not the most important event taking place on Randall's Island, New York City, on a sweltering July 11 that saw the temperature reach 97 degrees. Little Randall's Island, at the confluence of the East and Harlem Rivers and formerly the site of a potters' field and the Idiots' and Children's Hospital, had also attracted several giants of the New Deal: President Franklin D. Roosevelt, Governor Herbert H. Lehman, Mayor Fiorello H. La Guardia, Secretary of the Interior Harold L. Ickes, Works Progress Administration chief Harry L. Hopkins, Postmaster General James A. Farley, and New York City Parks Commissioner Robert Moses, among others. The first day of the Olympic trials happened to coincide with the dedication and opening of the Triborough Bridge, a Y-shaped span linking Manhattan, the Bronx, and Queens as well as both Randall's and Ward's islands.

Construction had begun seven years earlier, on October 25, 1929, the day the stock market crashed, and naturally soon thereafter slowed to a halt. But with the support of Ickes's Public Works Administration, Moses willed the bridge into existence. It was the first of his mammoth utopian projects — and less misguided than most of the others. Ickes had convinced the president that he should attend the opening ceremony, if only to thwart Moses's ego. "If you go to this affair," Ickes had said to

Roosevelt on June 23, "credit for the building of the bridge will go to your administration, where it belongs. Otherwise, Mr. Moses can be counted upon to bring glory to himself."

To celebrate the grand opening, thousands of New Yorkers lined the streets of Harlem as a parade of dignitaries made their way from midtown to the bridge. (Harlem, incidentally, needed a parade. Just three weeks earlier, the community had been plunged into a state of shock and depression when, across the Harlem River at Yankee Stadium, Max Schmeling had knocked out Joe Louis in the twelfth round of a fight Louis had been expected to win easily.) President Roosevelt left his townhouse on East 65th Street and, with the streets cleared of traffic, was driven up Fifth Avenue to 72nd Street, through Central Park to 110th Street, then northward on Seventh Avenue and finally onto 125th Street, Harlem's central thoroughfare. "Fully half of the men on the sidewalks were stripped to their undershirts and women wore as little as possible," the *New York Times* reported. "They gave the President an enthusiastic greeting."

Finally the caravan arrived at Randall's Island. Before the speeches began, Anthony Benedetto, a nine-year-old boy from nearby Astoria, Queens, sang for the crowd that had assembled. (Later, he became known as Tony Bennett.) When the singing ceased, Roosevelt spoke. Still fairly vigorous, the fifty-four-year-old chief executive stood supported at a lectern for the entirety of his brief speech. "Many of you who are here today," he said, "can remember that when you were boys and girls, the greater part of what are now the Bronx and Queens was cultivated as farmland. Not much more than one hundred years ago my own great-grandfather owned a farm in Harlem close to the Manhattan approach to this bridge."

Roosevelt went on to explain why the bridge and works like it were so needed. From his lectern, he could clearly see the hastily constructed stadium, just a few hundred yards away, where the Olympic trials were about to commence. Neither he nor any of the other speakers saw fit to mention the goings-on at the stadium, the lone structure on the 194-acre island, which had been cleared for the construction of the bridge.

More than seventy years after the Triborough Bridge made the island readily accessible, the stadium remains New York's primary venue for outdoor track-and-field events. But no event ever held there has led to more controversy than the Olympic trials that took place on July 11 and 12, 1936. The results of the 100-meter final in particular resonated for decades. Not because Jesse Owens won — that was almost a foregone conclusion — but because of the order in which the other sprinters finished and how those performances positioned them for the 4 x 100-meter relay race in Berlin three weeks later.

"This may be the best track and field meet ever staged in this country," John Kieran wrote in the *New York Times*, "and produce the best team the United States ever sent to Olympic Games." Even with Eulace Peacock hobbled, the sprints were expected to attract thousands of spectators.

At this point, American participation in Berlin was assured, yet some commentators were still fighting the boycott battle. It had become clear, especially to those few American writers who had covered the winter games in Garmisch-Partenkirchen, that the Nazis fully intended to use the Olympics in Berlin as a platform to promote their agenda. Brundage's promise to the contrary was dangerously naive. And so were all the statements made by the prospective Olympians, including Owens. As Walter White had passionately pointed out to him, at length, several months earlier, Americans had no business being in Berlin, regardless of the situation for Negroes in the United States and regardless of the laurels they might win for themselves in the fatherland.

The old lie that politics and sport were separate was betrayed by the Nazis' own publications. "Athletes and sport are the preparatory school of the political will in the service of the state," the political trainer Kurt Münch wrote in a Nazi-sanctioned book titled *Knowledge About Germany*. "Non-political, so-called neutral sportsmen are unthinkable in Hitler's state." More bluntly, Münch wrote that sports and politics in Germany could not be separated.

Curmudgeonly Westbrook Pegler addressed the issue of American hypocrisy on race issues. "The Nazis often point out that American Negroes are victims of discrimination," Pegler wrote in a column that appeared on the morning of the first day of the trials. "But Negroes are not barred from our Olympic teams. Many of them have worn the American shield in the past, and some of the most formidable athletes on this year's squad are colored." And again Pegler assailed the decision to keep the games in Germany and to allow the participation of American athletes. "Germany was awarded these games four years ago," he wrote. "But that was another Germany. At that time the international body which selected the site acted on the supposition that certain essential sporting conditions could be guaranteed this year. If the committee had had any idea that four years later the games would be used to ballyhoo Adolf Hitler and to endorse a regime of murder, persecution, and paganism, the program would have been awarded to another country."

If any of the hundreds of hopefuls at Randall's Island had good reason to heed Pegler's morning column, it was sprinter Marty Glickman, the eighteen-year-old son of Jewish Romanian immigrants. Glickman had just completed his freshman year at Syracuse University. He had also recently scored an impressive victory over Columbia's Ben Johnson in New York's Metropolitan Championships. Glickman had been born in the Bronx but grew up in Brooklyn, only a few miles south and east of Randall's Island. He had been both a track and a football star at James Madison High School, whose great rival, Erasmus Hall High School, featured at the time a Jewish quarterback named Sid Luckman. Glickman would eventually be enshrined in the Basketball Hall of Fame as a broadcaster, and Luckman would be enshrined in the Pro Football Hall of Fame as a player. After the Berlin Olympics, Glickman used his speed to tremendous effect for the Syracuse varsity football team as a running back, and then, after serving in the Pacific during World War II, segued into the burgeoning field of sports broadcasting. From the 1940s through the 1980s, he was one of the most widely respected and widely listened-to of all play-by-play announcers. But in 1936 he was just a poor Jewish kid trying to make the Olympic team.

Like almost everyone else at Randall's Island, Glickman stood in awe of Jesse Owens. He had advanced to the Olympic trials by defeating Johnson again at 100 meters in the eastern regionals in Boston. At Randall's Island, there would be two heats, and the top three finishers in each would advance to the final. A runoff comprising the four men who finished fourth and fifth in the two heats would determine the seventh finalist. Glickman was to run in the first heat, against Sam Stoller, of the University of Michigan, who was also Jewish and had competed frequently against Owens in the Big Ten; former Olympian Frank Wykoff; George Boone, of the University of Southern California; Billy Hopkins, of the University of Virginia; Ben Johnson; and Owens.

As the runners neared the starting line, John Kieran of the *New York Times* walked toward the east straightaway, where a crowd had gathered to see the Buckeye Bullet up close. "Which is Owens?" a man asked him.

"Wait a minute and you'll see," Kieran said, winking.

At the gun, Owens exploded from his crouch, charged to the front, and won easily, in 10.5 seconds. A few yards behind him, Sam Stoller finished second. Glickman was third. They would see each again in the final. Ben Johnson had pulled up lame at 60 meters and did not finish. In the other heat, Ralph Metcalfe won, Foy Draper placed, and Mack Robinson — whose younger brother Jackie would become the first black major-league baseball player of the twentieth century — showed. Sadly, Eulace Peacock all but limped home in last place. With his Olympic sprinting aspirations officially shattered, he would now concentrate on the broad jump. The seventh and final place in the 100-meter final went to Frank Wykoff, who had finished fourth in the heat won by Owens. In the runoff, Wykoff defeated Harvey Wallender, George Boone, and Edgar Mason, Jr.

The finalists drew black pills similar to marbles to determine their lane assignments. Mack Robinson was in lane 1, on the inside, Draper was in 2, Wykoff in 3, Owens in 4, Glickman in 5, Metcalfe in 6, and Stoller in 7. Three Californians, three midwesterners, and one Brooklynite. Three blacks, four whites. Five Christians, two Jews. The race, if nothing else, proved that in the United States, if not in Germany, virtually everyone had an opportunity to make the Olympic team. If the Germans had been paying attention — and they were — they would have noticed that five of the seven finalists in the most prestigious race in the Olympic trials of the leading track nation in the world would not have been permitted to train or compete in Germany. Only Wykoff and Draper would have been eligible to run in the fatherland, where the heterogeneity of the United States was viewed as a deadly sin.

For all seven men except Robinson, who was a 200-meter specialist, the 100-meter final was the culmination of several years of training and expectations. The first three finishers would represent the United States in Berlin and be in good position to win the Olympic gold medal, which they had good reason to believe would immortalize them. They all knew that it would be harder to succeed in this race than in the Olympic final — the field was much deeper. Still, if they didn't win, place, or show, they assumed they would help form an unbeatable 4 x 100-meter relay team. Lawson Robertson had indicated that the men who finished fourth, fifth, sixth, and seventh would make up the relay team.

The starter, Johnny McHugh, was a familiar figure to New York track fans — and to Glickman. Just after he called the finalists to their marks, McHugh noticed that Glickman was trembling so violently that he couldn't get his left leg into position. McHugh said, "Everybody up," and called Glickman over. "Walk a little, Marty," he said. "Jog up and back a bit, relax."

Three years younger and much less experienced than anyone else in the final, Glickman took a moment, walked around, then returned to his position in the fifth lane.

Jesse Owens, meanwhile, was perfectly placid in lane 4. It never occurred to him, not for a second, that he might not win. Even if he started poorly — now unlikely, thanks to Riley's and Snyder's relentless efforts — he knew that there would be enough time to recover. He still felt sorry for Peacock, but his rival's failure also somehow empowered him. Rather than thinking that what had happened to Peacock might happen to him, Owens allowed himself to think that somewhere a choice had been made by a higher power, between himself and Eulace — and he, Jesse, had been chosen.

Finally McHugh fired his pistol and they were off. Owens did not make a brilliant start, but Sam Stoller did. At 60 meters, the Michigan man was in the lead. Glickman was keeping pace with Owens, who was on his left, and Metcalfe, who was on his right. Suddenly, with about 30 meters to go, Owens and Metcalfe accelerated, "as if they were on a moving escalator at an airport and I was not," Glickman said. As their five rivals faded, Owens and Metcalfe rushed toward the tape. But Metcalfe could not keep up. Effortlessly, typically, Owens gained speed in the final 10 meters and broke the tape at full speed, in 10.4 seconds, 2 meters ahead of Metcalfe. Then came Glickman, Wykoff, Draper — they were bunched together — and then Stoller and Robinson.

The Owens-Metcalfe one-two was expected. Glickman's strong showing was not. When the judges lined up the top three finishers, they determined that Glickman had finished third. Ted Husing, the preeminent sports announcer of the time, interviewed Owens and Metcalfe and then said, "Now here's Marty Glickman, the kid from Brooklyn who finished third."

But within moments, Dean Cromwell had collared several of the officials. The man who was to coach the sprinters in Berlin argued that they had misjudged the order of finish. His Trojans, he told them, had been robbed. Wykoff was third, he said, and Draper fourth. Glickman, he insisted, had finished fifth. In the absence of photographic evidence, the officials were swayed and bumped Glickman all the way from third to fifth. Glickman later said that it was quite possible he had finished behind Wykoff — who was running his third 100-meter race of the day — but he was sure he had outrun Draper.

The reporters covering the trials took no note of the disputed finish, concentrating their efforts instead on finding new and better ways to praise Owens. To Associated Press sports editor Alan Gould, Owens was "machine-like." To Arthur Daley of the *Times*, Owens, clad in Ohio State's colors, was "a scarlet comet." In the *Chicago Tribune*, simplicity was prized. There, Owens was "the great colored athlete."

Owens's day did not end, though, with the 100 meters. There was still the broad jump, the event in which he was most dominant but, unlike the sprints, in which one or two errors might prevent him from qualifying for the Olympics. As Eulace Peacock limped and cleared only 23 feet, 3 inches — 3 feet less than his personal best — to finish ninth, Owens nonchalantly jumped 25 feet, 10¾ inches, a full 7 inches farther than the best effort of John Brooks, who finished second. Eddie Gordon, the defending Olympic champion, a black man from Iowa, was sixth. The 1932 gold medalist was no longer in Owens's league.

In the gleaming new locker room under the stands, Peacock mourned his career — he would never be an Olympian — and Owens looked ahead to Berlin. "I can't wait to get over there," he said. "You know, I've never been to Europe." Reminded that the 200-meter trials were looming the next day, he tried to summon a little bit of humility, but he could not even feign nervousness. "Well, really," he said, tearing off his shirt, "the two hundred I think is my strongest event. I know Mack will give me a good race, but that's my distance."

Owens was right. It was his distance, especially on July 12 at Randall's Island. He dominated the first heat, finishing in 21.2 seconds. Then, in the final, he decided to run really hard; he broke the tape in 21 seconds flat, equaling Ralph Metcalfe's un-official world record for 200 meters around a turn. Mack Robinson finished second and Bobby Packard of the University of Georgia third. Ralph Metcalfe, five times the national champion at 200 meters, wasn't quite at full strength on July 12. He barely qualified for the final, finishing third in his heat and then, in the final, fourth. It seemed he would compete only in the 100 meters in Berlin.

In fact, the United States would enter three men in each individual track-and-field event — for a total of fifty-one places — and forty-nine men qualified for those slots. Only one man qualified for more than one event — Jesse Owens, who had qualified for three, all events in which he held the world record, all events in which he would be favored to win the gold medal.

None of the reports of Owens's dominance mentioned the possibility that he might actually compete in four events and contend for four gold medals. Everyone assumed that Owens and Metcalfe would not run in the 4 x 100-meter relay, because they had qualified in the 100-meter sprint — and neither would Mack Robinson, because he had qualified in the 200-meter sprint. That meant that the 4 x 100-meter team would be made up of Wykoff, Draper, Glickman, and Stoller, who had finished third, fourth, fifth, and sixth, respectively, in the 100-meter final. Why would the American coaches choose not to include Owens and Metcalfe, their two best runners, on their relay team? Mostly because they had never needed to and rarely had. In 1932, for instance, none of the three best American sprinters — Eddie Tolan, who won the Olympic 100 meters; Ralph Metcalfe, who finished second; and George Simpson, who finished fourth — ran the relay, which the United States nevertheless won, in world-record time.

It was enough for Owens to have qualified in three events — enough for the writers, certainly, who expected him to equal Paavo Nurmi's 1924 Olympic trifecta, when he won three individual gold medals. Overall, including relays, Nurmi had won nine gold medals in the 1920s, at distances from 1500 meters to 10,000 meters. As the 1936 Olympic trials wrapped up, it was clear that Jesse Owens was the Nurmi of the sprints. "The great track and field meet closed as it had opened," John Kieran wrote, "with a lot of fellows chasing Jesse Owens and failing to catch him."

In the wake of the spectacular dominance of the black athletes at the Olympic trials, an effort was again afoot to explain their superiority. It had to be more than mere coincidence that black men won the 100 meters, the 200 meters, the 400 meters, the 800 meters, the broad jump, and the high jump. In both sprints, the broad jump, and the high jump, they finished one-two. "The Negro race's triumphs in every big track meet of the past few years are shockingly inconsistent with their relative number of performers," Shirley Povich (eventually a legend, but not yet) wrote in the *Washington Post* on July 13. "This department has always held its own theory concerning the sensational rise of colored athletes in track and field. Firstly, it is one of the sports in which they are permitted to contest against the best of the white race and thus the bulk of the Negro talent is directed into that line."

After the trials concluded, it would be three days before the U.S. Olympic team boarded the SS *Manhattan*, bound for Germany. In the meantime, the black Olympians remained at the Hotel Lincoln, a block off Broadway, with some of their white teammates. They could not stay with their other white teammates at the fancier hotels or at the still-restricted (to blacks and Jews) New York Athletic Club. During this lull, on the morning of July 14, a young columnist from Hearst's *New York American Journal* was working the Lincoln's lobby, sniffing out a story — he had to write seven columns a week — when he decided he would be better off knocking on Jesse Owens's door.

"Jesse, my name's Jimmy Cannon," the writer said when Owens opened the door to room 504. He extended his hand. "I write for the *Journal*. Mind if I ask you a few questions?"

"Go right ahead," Owens said.

Also not yet a legend himself, Cannon was completely un-interested in the Olympic trials per se. Like virtually everyone else, he had assumed that Owens would qualify easily in three events.

"What are you going to do in Berlin when the gun goes off?" he asked breathlessly.

"I am going to go as fast as my legs will take me," Owens said matter-of-factly. Then, somewhat more interestingly: "I think my time will depend on the boat ride. I have never been on a boat ride before, except to take a moonlight sail on the lake back home. But if I'm in good condition right now . . . if I'm in good condition, then — well, I'll be going good."

Cannon took note of the "little room" and the electric fans "wrinkling" Owens's loose blue polo shirt "with a hissing drone." "He seemed a little boy," Cannon wrote, "not a super-man who is the best in the business of running fast."

"Who are the guys you got to beat?" Cannon asked, wiping his brow with his handkerchief.

"The fellows on our team," Owens said, getting up from the bed on which he had been sitting. "They are the toughest compe-tition I got. The toughest. No doubt about that."

"Do you prefer to race against the best?"

"Yeah," he said, easing into the conversation, "I love competi-tion. It makes me run faster. It makes a track meet more interest-ing. That's what people like to see."

Then Cannon asked Owens about Ruth and how she handled all his traveling — and because Cannon had a lifelong weakness for Joe Louis, he asked about him, too.

"She gets a big kick out of me winning," Owens said. "She always sends me wires. I get a big kick out of that. You know, all that talk about marriage being bad for an athlete? That's not so. It's good for a fellow. It settled me down. I think that's good. No, I don't know where Joe Louis came from in Alabama."

Cannon wasn't Owens's only visitor that day at the Lincoln. Mark O'Hara of the *Daily Worker* also stopped by. He was mainly interested in drawing Owens into a political discussion, which Owens politely dodged.

"What do you think of James Ford running for vice president on the Communist ticket?" O'Hara asked. "He's the only Negro running for national political office."

"We're all Americans," Owens said, tucking his polo shirt into his dark slacks, "and if a Negro can win the position, he deserves it."

"What do you think of the Louis-Schmeling fight?" O'Hara continued, recognizing a dead end when he heard it.

"Joe Louis can whip him any day in the week. He wasn't in condition at the time."

Then O'Hara made another left turn, so to speak. "You know Paul Robeson will soon travel to the Soviet Union because he thinks it's the only place that there is complete equality of the races."

"It's a wonderful project there," Owens reportedly replied. "They're getting along." He was certainly unaware that Stalin's great purges were about to begin. The following month, in fact, the Soviet dictator put sixteen former party leaders on trial; all were found guilty and executed.

Then O'Hara said, "Would you care to go there?"

"I'd like to," Owens said, maintaining his politeness, "but time is valuable, and school starts in September."

The next day, in his syndicated column, Westbrook Pegler — who had continued to lend his acerbity to the boycott movement — predicted what would happen in Berlin once Owens, Metcalfe, John Woodruff, the 800-meter runner who was also black, and the others arrived. "The Nazis face a delicate problem in the track and field events," he wrote,

> where the Hitler youth doubtless will be defeated by a group of American Negro boys. When Max Schmeling came over to fight Joe Louis, Hitler, Goebbels, and the rest disowned him and his undertaking until he had won. It was then discov-

ered that he had been fighting for Hitler and the Nazi philosophy after all, and had proved the racial supremacy of the
Hitler youth. It was explained, too, that a Hitler youth naturally could not be beaten by a Negro. Perhaps after the Negro Olympians of the American team have beaten the Hitler
youth in the sprints and jumps the Hitler youth will turn out
to be Communists at heart.

No special intellectual suppleness was required to throw Max
Schmeling's victory back in the faces of the Nazis as evidence
of their disingenuousness. Expecting Louis to knock him out,
they had disowned the Black Uhlan. When instead he knocked
out Louis, the Nazis claimed him as a token of their superiority.
As politically unaware as he was, Jesse Owens still resented the
Nazis for making Louis's humiliation a propaganda tool. It had
been bad enough to listen on the radio as his friend and hero was
battered. It was worse to have his defeat presented as evidence
of black weakness. Millions of his fellow black Americans were
as upset as Owens. Now all the hopes that Louis had embodied
shifted to the Buckeye Bullet, who would carry those hopes with
him to Berlin, not as a burden but as an inspiration.

Owens's family lived in a three-room sharecropper's house like this one in Oakville, Alabama, when Jesse was a boy. *Allan Lebaron*

Jesse Owens of East Technical High School breaks the world record at 100 meters. The record was disallowed because of a strong tailwind. *New York Times Co./Getty Images*

Jesse, still in high school, shows his mother the four medals he won on June 17, 1933, at the national interscholastic track championships in Chicago. *Bettmann/CORBIS*

A student of form and his pupil: Charles Riley and Jesse Owens iron out the kinks.
Courtesy of the Ohio State University Archives

"My father's long, lion-spring legs": The captain of the Buckeyes poses with his coach, Larry Snyder. *Courtesy of the Ohio State University Archives*

A day to remember: Owens wins the 220-yard dash in world-record time in Ann Arbor on May 25, 1935. *Associated Press*

"Are we getting married tonight?": Jesse Owens takes Ruth Solomon as his lawful wedded wife. *Courtesy of the Ohio State University Archives*

On the eve of the boycott vote, Avery Brundage, president of the American Olympic Committee, stares down his nemesis, Jeremiah T. Mahoney, president of the Amateur Athletic Union. *Bettmann/CORBIS*

Lawson Robertson presides at the 1936 Olympic track-and-field trials on Randall's Island. *Associated Press*

Dean Cromwell of the University of Southern California coached thirteen world-record holders, and at every Olympics from 1912 through 1948 at least one of his charges won a gold medal. *Associated Press*

Defying the rules of shuffleboard, Owens poses for the cameras as the SS *Manhattan* prepares to leave New York Harbor. *Bettmann/CORBIS*

The youth of the world meets the German multitudes in the Olympic opening ceremony.
Courtesy of the Ohio State University Archives

Leni Riefenstahl always knew where to situate the cameras — especially those aimed at her.
IOC Olympic Museum/Allsport/Getty Images

The wages of sin: Olympic champion Eleanor Holm strikes a pose in 1934, two years before she and Avery Brundage clashed on the *Manhattan*. *Fox Photos/Getty Images*

"The voice, the hope, the prayer of the land": Adolf Hitler presides at his Olympics. *Fox Photos/Getty Images*

Ralph Metcalfe and Owens try to keep warm in the Berlin chill. *Fox Photos/Getty Images*

Owens breaks the tape in the 100-meter final in Berlin, a full step ahead of Ralph Metcalfe. *Associated Press*

Another victory for the "auxiliaries": Owens starts off perfectly in the 200-meter final. *Courtesy of the Ohio State University Archives*

Jesse Owens crashes into the pit after leaping 26 feet, 5½ inches in his final jump at the Olympics. *Courtesy of the Ohio State University Archives*

Fast friends: Luz Long and Jesse Owens exchange pleasantries on the infield of the Olympic stadium. *CORR/AFP/Getty Images*

The chosen ones: Jesse Owens, Ralph Metcalfe, Foy Draper, and Frank Wykoff, with baton, pose before the relay. *IOC Olympic Museum/Allsport/Getty Images*

Jesse Owens's name is chiseled on a stone wall at Berlin's Olympic stadium. *Courtesy of the Ohio State University Archives*

PART III

11

Olympia

IN TYPICALLY DRAMATIC FASHION, Leni Riefenstahl, the brilliant documentarian, was practicing high-jumping at the Grunewald Stadium in Berlin in the early spring of 1935 when she was approached by a middle-aged man in suit and tie.

"Fraulein Riefenstahl," he said, smiling, apparently unsurprised that someone who was not a high jumper was nevertheless high-jumping. "I've been planning to ambush you."

"An ambush," Riefenstahl said, wiping the sand from her legs. "What do you mean?"

"I have an idea," Professor Carl Diem said. "I'm supposed to prepare the Olympic Games in Berlin, and I would like to launch them with a huge torch race straight across Europe, from ancient Olympia in Greece to the new Olympia in Berlin. It will be a wonderful Olympics, and it would be a great pity if we couldn't record it on film. You are a great artist, you know a lot about sport. With your *Triumph of the Will*, you created a masterpiece, a film without a plot. You must make a film like that about the Olympics."

"Impossible," replied Riefenstahl, who was thirty-three but looked younger.

Undeterred, Diem, one of the two men most responsible for organizing the 1936 Olympics, persisted, and eventually the spectacle of the games and the challenge of filming them proved irresistible to Riefenstahl. "The possibility began taking shape," she later wrote.

> In my mind's eye, I could see the ancient ruins of the classical Olympic sites slowly emerging from patches of fog and the Greek temples and sculptures drifting by: Achilles and Aphrodite, Medusa and Zeus, Apollo and Paris, and then the discus thrower of Myron. I dreamed that this statue changed into a man of flesh and blood, gradually starting to swing the discus in slow motion. The sculptures turned into Greek temple dancers dissolving in flames, the Olympic fire igniting the torches to be carried from the Temple of Zeus to the modern Berlin of 1936 — a bridge from Antiquity to the present. That was my vision of the prologue to my *Olympia*.

Enraptured, Riefenstahl told Diem she would make the film.

Riefenstahl had achieved a measure of fame as a young woman in Germany in the 1920s, first as a dancer in the mold of Isadora Duncan, then as the striking star of several adventure films directed by her mentor, Dr. Arnold Fanck, in which she was usually seen scaling rugged Alpine peaks in shorts. In 1927, when the typically titled *Peaks of Destiny* was released in the United States, Mordaunt Hall, the *New York Times*'s film critic, wrote, "Riefenstahl is an actress with no little charm. Her overacting is to be blamed on Dr. Fanck." Hall obviously didn't know of Riefenstahl's capacity for self-indulgence. In 1932 she turned her talents to directing, emulating Fanck. Her first film, *The Blue Light*, an Alpine fairy tale, brought her to the attention of Hitler, who thought the film brilliant. Three years later, after *Triumph of the Will*, her propaganda film about the 1934 Nazi Party rally at Nuremberg, had been honored with several international prizes and her reputation had been made, Riefenstahl was pondering her next move.

Despite the success of *Triumph of the Will*, Riefenstahl had been quarreling with her boss, Joseph Goebbels, the Nazis' minister of propaganda and public enlightenment, whose romantic advances she had spurned. Goebbels was also jealous of her influence with Hitler. Riefenstahl had long been a favorite of Hitler's. "The point is that Hitler admires you," Rudolf Diels, the soon-to-be-deposed chief of the Gestapo, once told her, "and this arouses a great deal of envy and ill will."

Riefenstahl, who at the time was often referred to as the first lady of the Third Reich, flattered Hitler and flirted with him and believed that he would "save" Germany. Her first conversation with him about the Olympics film took place on Christmas Day, 1935, at Hitler's spartan residence in Munich.

"How did you spend Christmas Eve?" she asked.

"I had my chauffeur drive me around aimlessly," Hitler responded sadly, "along highways and through villages, until I became tired. I do that every Christmas Eve." He paused, then added, "I have no family and I am lonely."

"Why don't you get married?"

"Because it would be irresponsible of me to bind a woman in marriage," Hitler said. He was dressed casually, not in uniform. "What would she get from me? She would have to be alone most of the time. My love belongs wholly and only to my nation — and if I had children, what would become of them if fate should turn against me? I would then not have a single friend left, and my children would be bound to suffer humiliation and perhaps even die of starvation. I have tried to express my gratitude wherever I can, for gratitude is a virtue insufficiently valued. I have people at my side who helped me in my bad years. I will remain true to them, even if they do not have the abilities demanded by their positions."

Then Hitler said, "And what about you? What are your plans?"

"Hasn't Dr. Goebbels told you?" Riefenstahl asked. She had assumed that Hitler knew about the film.

When she told him, he was surprised. "That's an interesting challenge for you," he said. "But I thought you didn't want to make any more documentaries, that you only wanted to work as an actress."

"That's true," Riefenstahl said, "and this is definitely my last documentary. I thought it over for a long time. But I finally said yes because of the great opportunity that the IOC offered me, the wonderful contract with Tobis [the company that would help make and distribute the film], and, last but not least, the realization that we won't be having another Olympics in Germany for a long time." Hitler did not mention his plan to make Germany the permanent home of the Olympics.

She then explained, as best she could, the inherent difficulties of such a tremendous undertaking and her doubts about its eventual success.

"That's a mistake," Hitler said cheerfully. "You have to have a lot more self-confidence. What you do will be valuable, even if it remains incomplete in your eyes. Who else but you should make an Olympic film?" Then, to Riefenstahl's surprise, Hitler said, "I myself am not very interested in the games. I would rather stay away . . ."

"But why?" she asked.

Hitler took a moment, then said, "We have no chance of winning medals. The Americans will win most of the victories, and the Negroes will be their stars. I won't enjoy watching that. And then many foreigners will come who will reject National Socialism. There could be trouble."

At this point in their conversation, Hitler offered his requisite criticism of the Olympic stadium as inadequate to the scale of his ambition. It was too small. Too modest.

"But don't be discouraged," he said. "You are sure to make a good film."

Before they parted that night, Hitler offered his assessment of the character of Dr. Goebbels: "A man who laughs like that can't be bad." Then he took Riefenstahl down a corridor, unlocked a door, and showed her a bust of a girl decked out with flowers. "I told you why I will never marry, but this girl," he said, pointing at the bust, "is Geli, my niece. I loved her very much—she was the only woman I could have married. But fate was against it." (After finding a love letter from Eva Braun in Hitler's pocket, Geli had killed herself in the apartment in which Riefenstahl and Hitler were speaking—although there have long been rumors that she was murdered.)

Hitler had confided enough to Riefenstahl for one night. "Good luck with your work," he said as they parted. "You will not fail."

Hiring Riefenstahl to make the official Olympic film was both unsurprising and a stroke of genius. It was clear that she could be counted on to present to the world exactly what Hitler and Goebbels hoped to have presented. She would shamelessly romanticize the spectacle, even if it meant reenacting the important moments that her cameras failed to capture as they occurred. In fact, only German photographers handpicked by the organizing committee would be issued credentials, and only Riefenstahl would be allowed to film the games. "These arrangements," the *New York Times* correspondent in Berlin wrote presciently, "would appear to indicate clearly enough that the world at large is to see the Olympics through exclusively German lenses. It can safely be assumed that no opportunities for political propaganda that are likely to arise will be neglected and that untoward incidents from a propaganda point of view will be ignored." When, on April 25, the Olympic organizing committee announced that Riefenstahl would be directing the official film, she was identified in the *Times* as "a directress" who made mostly "propaganda films" for the Third Reich.

Riefenstahl did not dispute this characterization until after the Third Reich crumbled in 1945. In the meantime, she marshaled all of her considerable talents in service to making the documentary. It probably did not occur to her that the star of her film would be a handsome, dark-skinned non-Aryan whose grandparents had been slaves.

12

The Belle of the Ball

THE NIGHT BEFORE EMBARKING for Berlin, Jesse Owens found himself seated next to George Herman Ruth at a banquet honoring the American Olympic team.

"You gonna win at the Olympics, Jesse?" Ruth asked, slicing through his sirloin.

"Gonna try," Owens replied. He was nervous on two counts. Naturally, he was awed by the Babe, but additionally, watching the gargantuan Ruth eat, he was praying that nothing would be spilled on his rented tuxedo. It was obvious that coaching first base for the Dodgers was not a position that required physical fitness or table manners.

"Trying doesn't mean shit," Ruth said with a grin. "Everybody tries. I succeed. Wanna know why?"

Owens could only muster a nod.

"I hit sixty home runs a few years back because I *know* I'm going to hit a home run just about every time I swing that frigging bat," Ruth said. In his first full season of retirement, he could not quite bring himself to speak of his playing career in the past tense. Swallowing and continuing, he said, "I'm surprised when I *don't!* And that isn't all there is to it. Because I know it, the pitchers, *they* know it too. They're pretty sure I'm going to hit a homer every time."

Owens was duly impressed. Ruth was right. Confidence mattered, and when he boarded the SS *Manhattan* the following day, bound for Germany, he had never felt quite so confident.

Still, the voyage was a precarious one for him and his teammates, if only because of their proximity to Avery Brundage, who was on the lookout for revelers and anyone else who violated his code of conduct. They weren't exactly at close quarters with the AOC president; he and his cronies went first class, while the athletes were in steerage. Even in steerage, though, so much food was served that several Olympians ballooned as the ship sped toward Europe. Jesse Owens was lucky. His fine sense of balance made him particularly susceptible to seasickness, which made eating a less than pleasant pastime. He felt especially nauseated after the *Manhattan* encountered a tremendous storm on its second night at sea. A head cold also kept him in the stateroom he was sharing with Dave Albritton, Room 87 on Deck D, for some time.

"You've got to work your legs," Snyder told Owens, who was curled into a ball on his bed. "You can't just lie there for a week."

"Listen, Larry, if I so much as stand up, I'll be sick. There's no way I can run."

"Well, forget running," Snyder said. "You've got to eat. There won't be time to put the weight back on."

"I don't want to eat," Owens said. "I'll throw up."

Finally Snyder prevailed on him to go to the dining room. Before he knew what had happened, he was sitting down with three white southerners: hurdlers Glenn Hardin and Forrest "Spec" Towns and shot-putter Jack Torrance. Their accents and their jokes immediately made him feel ill at ease. This, he decided, was not the night on which he would come to terms with his boyhood in Alabama. Queasy already, he quickly excused himself and found a seat with Albritton.

"Did you hear about Eleanor Holm?" Albritton asked, negotiating some peas onto a fork.

"Hear what?" Owens said. Despite his illness, he cut a dashing figure in the dining room. He was wearing his handsome pinstriped suit—his only suit.

"She's been drinking with the writers," Albritton said, "and they're saying they might kick her off the team."

Owens pursed his lips and nodded his head. "I guess old Avery isn't messing around." While it was far from surprising to him that Eleanor Holm Jarrett had been partying—that surprised no one—he was disturbed by the thought that America's most decorated female swimmer might be prevented from competing. The daughter of a Brooklyn firefighter, Holm was only twenty-two years old, but she was sophisticated and experienced. In Los Angeles four years earlier, she'd won a gold medal in the backstroke. Her beauty and vivaciousness made her one of America's first female sports stars. They also attracted a famous suitor who became her first husband—the bandleader Art Jarrett.

"She won her place on the [1936] Olympic squad fresh from two solid years of trouping with her husband's band," Paul Gallico, one of her occasional drinking buddies, wrote, "during which time she sang nightly, never went to bed before three or four in the morning, and carried her share of the burden of the traveling cabaret show business by drinking with the local big shots and helping her husband sit up with them until the waiters whipped the cloths from the table and turned out the lights."

If there was one man who was impervious to Holm's charms, it was Brundage, who found her behavior repugnant, an embarrassment to the American Olympic team. Disregarding the directives of the AOC, the swimmer brazenly consorted with writers and other passengers aboard the *Manhattan*. One night early in the voyage, she outdid herself, in the company of Charles MacArthur, the gifted playwright, whose wife, the equally gifted Helen Hayes, had turned in relatively early. Holm, whose fondness for champagne was self-confessed and well known, drank enough of it with MacArthur to make her subsequent unsteadiness quite conspicuous. Certainly the Olympic officials she liter-ally bumped into noticed.

This party, during which Holm declined to heed hints that she would be well advised to return to the athletes' quarters, cul-minated in the American Olympic Committee's serving a sharp warning that put the backstroke star on probation with "one more chance," in Avery Brundage's words.

"If I was her," Owens said to Albritton, "I would stay in my room."

"Doesn't matter now," Albritton said, as he speared another pea. "She's a goner. They'll find a way to get rid of her."

On the final night of the crossing, Holm would once again get drunk, this time within full view of an official chaperone. The next morning, as the *Manhattan* docked in Bremerhaven, Brund-age kicked her off the team, igniting a storm of criticism.

"The campaign of vindictiveness against the little lady was inspired not so much by the fact that she drank champagne," Gallico wrote later in her defense, "but that she could drink and still win. That was and always is the unforgivable sin." Then, taking aim at Brundage and his cronies, he continued, "The Am-ateur fathers were always delighted to avail themselves of the girl's services — gratis, of course — at their swimming meets and cash in on her attractions; but when she punctured an illusion for them, they repaid her with public disgrace. It was a noble ganging-up."

Lawson Robertson was among those ganging up on Holm. In fact, he didn't think women should participate in the Olympics, period. "The Greeks had the right idea over 2000 years ago," he said, "when all women were barred." That comment, which to-day would be grounds for immediate dismissal, drew hardly a sniff. For many of the titans of the Olympic movement in 1936, the presence of women on their team was as objectionable as the presence of blacks and Jews. In other words, such athletes were tolerated, barely, and far from equal.

Most of the preeminent sportswriters of the era were not what could be called liberal thinkers, but compared to the Olympic solons, they were proto-integrationists and proto-feminists. Their mixed praise for the blacks and women was better than nothing.

Comparing Holm to other great athletes known for the laxness of their training regimens — including the great middleweight Harry Greb and the golfer Walter Hagen — Grantland Rice wrote that she "has been one of the few immortals who break training rules and world's records together." Rice defended her, sort of, writing that she had trained the same way for her first Olympics, in Amsterdam in 1928, when she was fourteen years old. It is unclear whether he meant she had been drinking then too.

"I have always taken a drink when I felt like it, when I felt I needed it to keep relaxed," Holm told Rice. "This trip has been no different from any other I have taken, including those to the Olympics at Amsterdam and Los Angeles. I am still in condition to swim backstroke in record time. Facing my third Olympics, with the records I have made, I still feel I am the best judge of what I should do, within reason. I was quoted as saying before the Olympic trials that I had trained on champagne and cigarettes. I did not mean this literally. I merely meant that I had led my natural life to keep from going stale from so much competition."

As usual, Westbrook Pegler saw clearly that Brundage was a buffoon who could always be trusted to make the wrong decision. "He is very sensitive to personal flattery and honors at the hands of the soft-soap department of foreign governments," Pegler wrote, "and undoubtedly regards his election to the International Olympic Committee as recognition of his sporting idealism rather than his susceptible innocence."

If nothing else, it can be said of Brundage that he cared not a whit about public relations — which was what scared coaches such as Larry Snyder and athletes such as Jesse Owens. If there was an unpopular, reactionary stance to be taken, Brundage took it. In the case of Eleanor Holm Jarrett, he was almost universally criticized. As might be expected, the *Daily Worker* took shots at him. "The smug gentlemen of the Olympic Committee knew all about Eleanor before she was given a place on the team," Fred Farrell, the *Daily Worker*'s sports columnist wrote. "They knew they could just as easily keep a bird from flying as Mrs. Jarrett from drinking. Then, why they should get in a terribly nervous state because the young lady became crocked I don't know."

Eventually even Hitler came to Holm's defense. She later claimed that when she met the Führer during the games, he told her, through his interpreter, that if she had been on the German team, he would have waited until after the Olympics to punish her — but only if she had lost. "Hitler asked me himself if I got drunk," she said. "He seemed very interested, and I said no." (The Nazi leaders were quite taken with America's disgraced sweetheart. Both Göring and Goebbels flirted with her, and Göring even gave her a sterling silver swastika. "I had a mold made of it," she said, "and I put a diamond Star of David in the middle of it.")

Unlike most of his teammates, who were gossiping as they gained weight, Jesse Owens remained relatively uninterested in the saga of Eleanor Holm Jarrett. When Albritton tried to tell him the most recent developments, he told him to be quiet. "All I'm thinking about is taking home one or two gold medals," he said one night on his bed, all but sticking his fingers in his ears. Owens's goals for the remainder of the voyage were simple. He wanted to avoid rich foods, stay in shape, and steer clear of Brundage.

However, he did find himself in the middle of a different, somewhat less sordid controversy. As the *Manhattan* sliced through the North Atlantic, Larry Snyder was fighting Robertson and Cromwell to maintain control of Owens's training. At first it had seemed that Robertson himself would coach the sprinters. Then he delegated the assignment to Cromwell. But it was still unclear who would coach the best athlete on the team. Would it be an official American track coach or his personal coach? Finally Robertson went to Owens and asked him, on the record, which coach he wanted. "It's up to you, Jesse," Robertson said. They were at the rail, and the ship was pitching slightly in a gale. "You know Dean is one of the most respected coaches in the world." The implication — that Snyder was not — was clear.

"Larry, of course." It was chilly, and Owens was in no mood to be polite. His tone was dismissive. There was never any chance that he would sign off on the attempted Robertson-Cromwell coup.

"Fine," Robertson said, clapping his hands together as if to say, *Have it your way, you foolish young man.* He looked at Owens coldly, telegraphing his disapproval. "Snyder it is." Robertson did not again attempt to separate Owens and Snyder.

Unlike Robertson and Cromwell, Snyder was making the trip purely as a civilian, paying his own way. He was called an associate coach, but for him there were no uniforms or identity cards. He would have to finagle his way through the Olympics. In other words, the man responsible for the training of the American Olympic Committee's greatest hope was officially an interloper.

Of course, when Owens told him about Robertson's attempt to usurp him, Snyder was peeved — and he did not care who knew it. "I am willing to cooperate as far as possible," he told reporters on the starlit evening of July 23, as the *Manhattan*, which had stopped earlier in the day in Plymouth, England, plied the North Sea, "and surely will handle Owens, Dave Albritton, and any others I'm requested to coach. But I dislike being classified as a knot-holer [knot-holing, now a largely lost art, is watching a game, usually a baseball game, through a hole in the fence]. I have been given an official runaround, but the important thing to me is to get Owens into the best condition to capture the triple Olympic honors, which he is sure to do."

Snyder was just building up a head of steam when Alan Gould of the Associated Press asked him to respond to comments Robertson had made that were critical of Owens's starting technique. Robertson, Gould said, had used the phrase "slowness off the marks" to describe Owens's weakness. Now Snyder was truly indignant. His head was filled with well-founded visions of Robertson and Cromwell conspiring to undermine him. It was bad enough that he had to endure their condescension at the meets with USC and the Penn Relays. It was bad enough that they had made it clear that they considered him little more than Owens's caddy. Now they were questioning his competence. That they were the unrivaled giants of amateur track and field only made it worse. This was Robertson's fourth consecutive trip to the Olympics as the head coach of the American track team. Before that he had served as an assistant coach, in 1912 and 1920. He had competed at the Olympics, in the standing jumps, in 1904 and 1906. Then there was Cromwell, who had more medal hopefuls on the boat than all the other track coaches combined.

"You can tell Mr. Lawson Robertson that Jesse Owens can beat any sprinter at any distance," Snyder said. Then, measuring his words precisely, he added, "There's no payoff at fifty meters in a hundred-meter race. Owens is no jackrabbit, but the smoothest piece of running machinery we have ever seen."

13

The Battle Tent of
Some Great Emperor

As Jesse Owens and his teammates were still at sea, the Third Reich was rolling out its red-and-black carpet for international celebrities, including America's most famous citizen, Charles Lindbergh. Visits such as those Lindbergh made were the raison d'être of the Berlin games. As much as it meant to the Germans to prove that they were athletically capable, it meant much more to them to show off their new society — minus, of course, its uglier idiosyncrasies.

It had been nine years since Lucky Lindy had touched down at Le Bourget, outside Paris, in the *Spirit of St. Louis,* completing the first solo flight across the Atlantic Ocean. It's difficult now to appreciate the magnitude of his achievement and his subsequent deification. He was handsome, wholesome, and heroic. He was celebrated in poetry and song. In the words of William L. Shirer, who was constitutionally averse to hyperbole, Lindbergh was "the idol of the nation, its most publicized citizen, its one authentic hero."

Shirer had covered Lindbergh's landing at Le Bourget and had followed, with the rest of the world, the trajectory of his life and career. He read with horror of the kidnapping and murder of Lindbergh's infant son. He reacted with a different kind of horror when he learned that Lindbergh and his wife, Ann Morrow, had moved in 1935 to an island off the coast of Brittany, to be near the island's best-known resident, Dr. Alexis Carrel, a French-American Nobel laureate in medicine with an affinity for fascism. Carrel, incidentally, was the first surgeon—and one of only three ever—to win the Nobel Prize for medicine.

In Berlin, Lindbergh declined to meet formally with American reporters—he had grown extremely wary of the press—choosing instead to spend most of his time in the company of his fellow aviator Hermann Göring, "the fat chief of the rapidly growing German air force," as Shirer described him. Göring and Lindbergh spent a pleasant afternoon switching off at the controls of the *Field Marshal von Hindenburg*—not the famous dirigible but an enormous eight-engine plane built by Lufthansa. Lindbergh did make a speech decrying the development of bombers that, he predicted, might soon lay waste to all of Europe. But when Shirer and a few other American correspondents based in Berlin heard that he had been won over by Göring, Goebbels, and their ilk, they cornered him at one of Goebbels's parties and tried to disabuse him of his impressions—to no avail. "Lindbergh proceeded to tell us what the situation was in Germany," Shirer wrote, "and to express his admiration for what had been achieved here. He, too [like so many prominent visitors to Germany at the time of the Olympics], had found a happy, united people, he said. As an airman, he was particularly impressed by the German air force and the progress of German aviation in general. Once again, we did not get a word in."

Like most reporters who had been stationed in Germany during the Nazis' rise to power, Shirer found the Third Reich utterly contemptible. But the Nazi propagandists were talented, and in an uncertain world, Charles Lindbergh was far from the only foreigner to fall under their sway. Still, there were those capable of seeing the regime for what it truly was.

One of the more skeptical visitors to Berlin as the Olympics were about to commence was a thirty-five-year-old writer from

Asheville, North Carolina. He was a drunk and a bigot, but neither of these qualities hindered his career as a novelist. In fact,
by the time tuberculosis killed him at the age of thirty-seven, he
had achieved a measure of fame so profound that nearly seventy
years later his novels are still widely read and admired. An in-
nocent abroad, Thomas Wolfe went to Berlin to cover the games
and to see for himself what Hitler was up to. He was also taking
notes for a novel that would become *You Can't Go Home Again*,
which was published posthumously in 1940. In it, George Webber, Wolfe's alter ego, flits from New York to Paris to Berlin at the
time of the Olympics. Like Wolfe, George Webber is both awed
and terrified by Hitler's capital. Perhaps more than any other
American who attended the games, Wolfe was able to communicate the prevailing mood:

> George observed that the organizing genius of the German
> people, which has been used so often to such noble purpose,
> was now more thrillingly displayed than he had ever seen it
> before. The sheer pageantry of the occasion was overwhelm-
> ing, so much so that he began to feel oppressed by it. There
> seemed to be something ominous about it. One sensed a stu-
> pendous concentration of effort, a tremendous drawing to-
> gether and ordering in the vast collective power of the whole
> land. And the thing that made it seem ominous was that it
> so evidently went beyond what the games themselves de-
> manded.

Although the Germans had packed their uniforms into closets
and their anti-Semitic posters into warehouses, they could not
hide what was in their hearts. And the thought of it made Wolfe
shiver. It was clear that a great society was building its strength
and flaunting its resourcefulness and that there would be dire
consequences for anything and anyone that stood in its way.

To its Olympic visitors, Berlin was both sinister and majes-
tic. Maybe it wasn't quite Vienna or Paris yet, but it seemed to
be only a matter of time before it would match and then surpass

all its rivals. Despite the official decrees mandating civility, the entire city was on an aggressive footing, jutting out its chin, daring someone to take a shot. The populace, on the whole, was still enthralled with Nazism — especially after the events of March, when the Rhineland had been remilitarized and France had done nothing to counter that action. "Hitler has got away with it!" Shirer had marveled in his diary. Unrestrained enthusiasm floated through the city, which was decked out in festive Nazi banners. "From one end of the city to the other," Wolfe wrote, "from the Lustgarten to the Brandenburger Tor, along the whole broad street of Unter den Linden, through the vast avenues of the faery Tiergarten, and out through the western part of Berlin to the very portals of the stadium, the whole town was a thrilling pageantry of royal banners — not merely endless miles of looped-up bunting, but banners fifty feet in height, such as might have graced the battle tent of some great emperor."

Wolfe could not have known — few people did — that more than grime had been scrubbed from Berlin's streets in the days leading up to the games. On July 16, in compliance with an order from the Interior Ministry, Berlin police had rounded up about eight hundred Gypsies and deposited them in a filthy, verminous camp just outside the city, in Marzahn. A week later, Richard Walther Darré, the Argentine-born German minister of food and agriculture, issued a decree that was sent to local authorities throughout Germany to enforce: "All anti-Semitic posters must be suppressed during the period in question. The fundamental attitude of the Government does not change, but Jews will be treated as correctly as Aryans at this time . . . Houses on the main roads must be whitened, and even repainted if possible. Street lighting must be improved. Streets and squares must be cleaned. Agricultural workers in the fields must not take their meals near the roads, nor pass near the roads."

Darré later made a name for himself in the field of eugenics. As the head of the SS Race and Resettlement Office, he wrote, "Just as we breed our Hanoverian horses using a few pure stallions and mares, so we will once again breed pure Nordic Germans." He was among the architects of what came to be known as the Final Solution, which led to his conviction at Nuremberg (and a stunningly brief five-year prison sentence).

In the summer of 1936, though, Darré wanted the Third Reich to appear to be inclusive—at least for a few weeks. The Nazis declared a ceasefire against virtually all their traditional targets. "No action against unruly Catholics and Protestants," Shirer wrote. "No savage attacks against the 'decadent' Western democracies and 'Jewish-dominated' America. All was, for the moment, sweetness and light, except for an occasional diatribe against the terrible Bolsheviks."

Under Goebbels's direction, the Reich Press Chamber issued several decrees to domestic news outlets in the weeks leading up to the opening ceremony. Even as the Germans bowed to international pressure by allowing the fencing champion Helene Mayer to compete in the games, Goebbels's office made it clear that her Jewishness (actually, she was half Jewish) was a taboo topic. Back in February, the press chamber had announced that "no comments should be made regarding Helene Mayer's non-Aryan ancestry or her expectations for a gold medal at the Olympics." On July 16, just in case anyone had forgotten, the chamber decreed that "press coverage should not mention that there are two non-Aryans among the women: Helene Mayer and Gretel Bergmann." Bergmann, however, was dropped from the German team just before the games, despite the fact that she held the national record in the high jump.

The Nazis' efforts to placate the international community were convincing only to those who wished to be convinced. Perhaps no one saw their true intentions more clearly than Fred Farrell of the *Daily Worker*. Of all the distinguished reporters and columnists who were writing about the Olympics as the opening ceremony neared, Farrell had the best sense of what was at stake—the prestige of the Third Reich. Frequently the *Daily Worker*'s stories were tainted by the paper's insistence on seeing everything through the prism of its politics. In the case of the 1936 Olympics, though, its prism was perfectly focused.

In an impressive display of prescience, Farrell wrote, "I have an idea that these Olympic Games ... will have repercussions that will not help toward the amity of nations as they are supposed to do. There will be dark trouble clouds along the horizon. They may develop into thunderstorms that will make Eleanor Holm Jarrett's champagne party look like an afternoon tea for the kiddies." Then he anticipated the controversy that lingers even now. "In the long run," he wrote,

> I'm afraid that noblest sportsman of them all, Herr Hitler, will find the Olympic Games the bitterest disappointment. When the time comes to present medals to Negroes and Jews his ideas of racial supremacy for the people he wants to be supreme will suffer something of a shock. Will he shake hands with Jesse Owens, for example? The Reichsführer will be on the spot — but definitely. If he doesn't take Jesse's hand he will be marked as a bad sportsman. If he does how can he laugh it off to his stormtruppen who have been taught that Negroes and Jews have no place in the Hitler scheme of things?

For the most part, however, the German propaganda machine was remarkably successful at portraying the regime in the most favorable light. Still too weak to state his true intentions, Hitler was forced to pretend to want nothing more than peace and a place of honor for Germany in European affairs. He needed time. Time to build strength. Time to perfect his plans. Time to rearm fully. And if he did not lull his neighbors into a false sense of at least partial security, he would not be afforded the time he so desperately needed. For Hitler and his colleagues, the Berlin games were the ultimate opportunity not to promote their agenda but to hide their agenda under a cloak of hospitality, prosperity, and efficiency. "This is the most loathsome feature of the swastika crusade," Victor Klemperer, a fifty-five-year-old, Jewish-born history professor wrote in his diary in Dresden, "that it is conducted hypocritically and in secret. 'We' are not conducting a crusade, 'we' do not shed blood either, we are completely peaceable people and only want to be left in peace! And at the same time not the smallest opportunity for propaganda is missed."

Two weeks before the opening ceremony, Paul Gallico arrived in Berlin. He was eager to get acclimated to the site of the Olympics — eager too to be as far as possible from Chicago, where his twenty-one-year-old second wife, Elaine, was busy testifying against him in divorce proceedings. As she was accusing him of assault and lesser misdeeds, Gallico was walking the grounds of the sparkling athletes' village the Germans had so carefully planned and built. Gallico, who had fought the Germans as a navy gunner in World War I, was surprised to realize that his former enemies had situated the village — which was already housing the youth of the world, who were gathering ostensibly to promote international understanding — next to an army shooting range. The juxtaposition was too delicious to ignore. "Your wandering correspondent," he wrote, "hopes he has not spoiled anything for the visiting firemen who will be arriving here shortly and who might find good copy in the fact that the anxious Germans are rehearsing for the next war right next door to where the athletes are flexing their muscles and practicing to win the great peace games of 1936."

Unlike Gallico, though, most observers raved about the athletes' village, despite its location on the outskirts of the vast capital. Beautifully landscaped and luxuriously appointed, the Germans' tidy brick athletic enclave put the bungalows of the Los Angeles Olympics to shame. It was staffed by dozens of bi- and multilingual volunteers, and the chefs in residence prepared high-quality food familiar to delegations from Japan to Peru. Athletes were encouraged to mingle with one another and with visitors in common areas. Open to the public, the village became a popular destination for tourists before and during the Olympic fortnight — a situation unimaginable after the massacre of Israeli athletes in Munich in 1972.

The gifted black athletes from the United States who would soon occupy the village naturally were the chief topic for America's black press. As the opening ceremony neared, papers such as the *Chicago Defender* and the *Amsterdam News* all but dropped every other story they had been covering. After all, sixteen black men and two black women would represent the United States (the American team in Los Angeles had included only six blacks), and in the wake of Joe Louis's knockout at the hands of Max Schmeling, Louis was a source of embarrassment, not pride. R. Walter Merguson of the *Pittsburgh Courier,* one of a handful of black reporters in Berlin, was typically breathless: "Jesse Owens and his brown-skinned brothers of the United States are the 'men of the hour' in this ancient capital of a mighty nation today as the 'zero hour' for this Eleventh Olympiad approaches. On every tongue, the name of the fleet-footed sons and daughters of Ham are being bandied about."

On June 24, 1936, the SS *Manhattan* docked at Bremerhaven on the North Sea. Within a few hours, Jesse Owens and his teammates — as well as Larry Snyder — were in Berlin, marching in a parade down Unter den Linden. After an audience with the mayor, the Americans were taken to their quarters 15 miles away at the Olympic village. They were greeted by members of the Hitler Youth.

Snyder stayed behind at his hotel in Berlin, but Owens and Dave Albritton threw their suitcases on their beds and headed out to find the track where they would train. After nine days aboard ship, Owens was eager to shake off his wobbly sea legs.

"You know, I feel great," he said to Albritton as they jogged down the cinder track at the edge of the village. "I don't think I could feel any better."

"Take it easy, Jesse," Albritton said, although usually Owens was telling *him* to take it easy. "Don't rush it."

"Dave," Jesse said, "don't worry about me. I think this place suits me."

Owens embraced everything about the village. Here, in the heart of the Third Reich, he found a utopian athletes' paradise. He loved the food and the companionship — probably in that order. He wrote in his diary that he was eating "steaks, and plenty of it, as well as bacon, eggs, ham, fruit, and juices." And for a good time, he gravitated to the easygoing Australians. "What liars they are!" he noted in his diary, admiringly.

14

The Youth of the World

BERLIN: SATURDAY, AUGUST 1, 1936

W HEN IT WAS MADE CLEAR to Leni Riefenstahl that there would be no room for her cameras in the Führer's box at the opening ceremony of the games of the Eleventh Olympiad, she did what she always did in such situations. She marched into the Reichschancellory and requested help from the Führer himself.

"You can put your cameras there," Hitler said distractedly. Even with his blizzard of appointments and appearances, he usually found time to see his favorite director, but he was not as affable as usual. "I'll approve it. They'll only be in the way for a few minutes."

"Thank you," Riefenstahl said, heaving a sigh of relief (as she later recalled).

"You know, Leni," Hitler continued, echoing the sentiments he had expressed on Christmas day, "I'll be glad when the entire Olympic commotion is over. I'd much rather not visit the games at all."

Riefenstahl was dumbstruck—all the more so when she realized that the Führer was also uninterested in her film. Still, she walked out of the chancellory satisfied—armed with the Führer's permission to do as she pleased.

When August 1 finally arrived, Riefenstahl went to the stadium at dawn, dodging the steady rain, fretting over every camera position and angle, barking orders to her cameramen, preparing to bark at the still cameramen, who she sensed would ruin her shots by impudently trespassing on her space. She was exhausted but well dressed. A fashion reporter covering the opening ceremony wrote in the *Los Angeles Times* that Riefenstahl was wearing "a smart gray flannel culotte suit."

Meanwhile, out on the streets of the city, now teeming with visitors from around the world, excitement was growing. Five years after the games had been awarded to a very different Germany, they were finally about to begin. Never before had the capital experienced anything like this—not even in 1871, when Bismarck had united the country. This was different. Now the whole world was watching, and the Germans were prepared to give it the greatest show ever staged, certainly the greatest Olympics.

As the countdown to the lighting of the Olympic caldron continued, the athletes made their way from the village to the stadium (the male athletes, anyway; the women were staying in a dorm adjacent to the stadium). Spectators began filing in hours before the ceremony was to begin, and more crowds were on the streets, behind barriers, hoping to catch glimpses of all the athletes and officials and celebrities who were motoring to the ceremony. As much as they wanted to see Karl Hein and Luz Long, their brilliant hammer thrower and broad jumper, respectively, there was one man above all others they waited for. Only when their anticipation became a desperate hunger would he satiate them with his appearance. Only when the crowds were frothing with excitement would he make his triumphal approach.

Finally, hours after assembling, the crowds lining the streets sensed that their Führer was approaching. Their excitement, undimmed by the gray skies, now reached fever pitch—as it would throughout the games whenever he was about to appear. They would line the route to the stadium, hoping merely to catch sight of him as he came and went each day. Thomas Wolfe was among the throngs and wrote, "From noon till night they waited for just two brief and golden moments of the day: the moment when the Leader went out to the stadium, and the moment when he returned." When the Führer could at last be seen, the crowd surged and gasped. "At last he came," Wolfe wrote, "and something like a wind across a field of grass was shaken through that crowd, and from afar the tide rolled up with him, and in it was the voice, the hope, the prayer of the land."

Now, as Hitler neared the stadium, a frenzy transformed the crowd into a shrieking, writhing, almost inhuman mass. Berlin was suddenly Nuremberg — festooned with black-and-red banners, populated by hysterical Nazis. For Grantland Rice, the scene was undeniably ominous. "Just twenty-two years ago this day the world went to war," he reminded his millions of readers. "On the twenty-second anniversary of the outbreak of that great conflict I passed through more than 700,000 uniforms on my way to the Olympic Stadium — brown shirts, black guards, gray-green waves of regular army men and marines — seven massed military miles rivaling the mobilization of August 1, 1914. The opening ceremonies of the eleventh Olympiad, with mile upon mile, wave upon wave of a uniformed pageant, looked more like two world wars than the Olympic Games." Apparently the effort to tone down the martial atmosphere that had been apparent in Garmisch-Partenkirchen had not yet succeeded.

Inside the stadium, the crowd of more than 100,000 could sense that the Chancellor was nearing. First they saw the gigantic dirigible *Hindenburg* fly by. Then, Frederick T. Birchall wrote, "from far away, a sound of cheering and a fanfare of distant trumpets, a sound ever growing nearer." Hitler was close.

The crowd waited as the Olympic hymn was conducted by its eminent composer, Richard Strauss. ("To Strauss the composer I take off my hat," Toscanini once said. "To Strauss the man I put it back on again.") After the hymn, cannons were fired and pigeons released. And then, at four o'clock, with the crowd hushed, Hitler presented himself at the west portal. He was wearing a plain brown uniform and a peaked cap. Flanked by

Lewald and De Baillet-Latour—both dressed like World War I–vintage statesmen in dark tails and starched wing collars, with heavy medals around their necks—the Chancellor projected an unmistakable air of modernity. Here was a man for his time, not a spiked-helmeted Prussian weighed down by decorations, not a penguin-suited diplomat in black and white, but a prophet of the new Germany, the creator of a thousand-year reich. Together, Hitler's entourage marched across the field as a hundred thousand "Heils" bounced against each other in the muggy bowl. "At his coming," Birchall wrote, "these assembled thousands rose to their feet, with their arms outstretched and voices raised in a frantic greeting."

The musical accompaniment to this procession was, of course, Wagner, whose most martial strains blared through the stadium loudspeakers. The music faded when Hitler and his small attachment reached the honor stand, where Professor Carl Diem's four-year-old daughter, in a blue dress, ran out to greet them. Her blond hair was tied together by a chaplet of flowers. She saluted the Führer, curtseyed, and offered him a nosegay. He touched her hair, spoke softly to her, and accepted her gift. Then he turned around to acknowledge the cheering multitude.

Meanwhile, Leni Riefenstahl's army of sixty cameramen was exposing tens of thousands of feet of film in an effort to capture everything that was going on in the stadium. Two cameramen had been positioned on the rostrum where Hitler and the loftiest dignitaries were ensconced for the ceremony, but their equipment was unsightly, and Goebbels ordered the cameras removed. Riefenstahl refused.

"Have you gone mad?" Goebbels screamed as Riefenstahl blocked two SS men from the camera position. "You can't stay here! You're destroying the whole ceremonial tableau. Get yourself and your cameras out of here immediately!"

Quivering with rage and frustration, Riefenstahl broke into tears. "Herr Minister," she said, "I asked the Führer for permission way ahead of time—and I received it. There is no other place from which we can film the opening address. This is a historic ceremony — it cannot be left out of an Olympic film."

Unmoved by this entreaty, Goebbels continued to shout. "Why didn't you set your cameras on the other side of the stadium?" he bellowed.

"That's technically impossible!" Riefenstahl countered. "The distance is too great."

"Why didn't you build a tower next to the rostrum?"

"They wouldn't let me."

Still furious, Goebbels barely acknowledged Hermann Göring as he made his way to his seat. In defiance of conventional wisdom, the general's all-white uniform actually made him seem less obese. Intensely distrustful and disdainful of the propaganda minister, Göring raised his gloved hand, as if to say, *Silence*, and Goebbels obeyed.

"C'mon, my girl, stop crying," Göring said in his most consoling tone. "There's room here even for my belly."

Down on the field, as the old men and the younger man and the little girl walked together to the Chancellor's box, a band struck up "Deutschland Über Alles" and then, of course, "The Horst Wessel Song," Nazism's anthem to one of its martyrs, a pimp who had been murdered by Communists. The rain stopped falling, and a tremendous gong was heard from the highest platform in the stadium. The Olympic bell, on which the words "I summon the youth of the world" had been inscribed, was pealing, signaling the commencement of the parade of nations. The Greeks, as always, led the procession, followed in alphabetical order by all the nations participating — except the Germans, who would be last. One by one, the athletes and officials marched by the reviewing stand, their flag-bearers in front.

The question in thousands of minds was which nations would give the Nazi salute as they passed Hitler and which would give the Olympic salute. The spectators took careful note. Naturally, Nazi salutes were greeted with cheers and Olympic salutes mostly indifferently. Complicating the situation was the similarity of the two salutes. The Nazis saluted by jutting out their right arms directly in front of their bodies, while in the Olympic salute the right arm was presented at a slight angle to the right. Still, in mass formations, the crowd generally could distinguish between the two.

When the Afghans, Bermudans, Bolivians, Icelanders, and of course the Italians honored Hitler by holding their right arms stiffly in front of them, the crowd cheered. But no one could match the obsequiousness of the Bulgarians. Not only did they perform perfect Nazi salutes, they went a step further, by breaking into a goose-step in front of the Führer. The Chinese and the Filipinos, in contrast, did not use either salute; instead they put their hands on their hearts. The New Zealanders mistook a German athlete clad in white for Hitler and removed their hats when they passed *him*. Oddly enough, the French received the most thunderous applause, despite their Olympic salutes — the Germans were probably still feeling grateful for French indifference to their reoccupation of the Rhineland, although it's also possible that they simply thought the Olympic salutes were Nazi salutes. In any event, Albert Speer detected something profound in the crowd's reception of the French delegation. The French "had marched past Hitler with raised arms and thereby sent the crowd into transports of enthusiasm," he wrote. "But in the prolonged applause Hitler sensed a popular mood, a longing for peace and reconciliation with Germany's western neighbor. If I am correctly interpreting Hitler's expression at the time, he was more disturbed than pleased by the Berliners' cheers."

The Germans also loudly cheered their Austrian cousins, whose national ambivalence on the subject of Nazism was reflected in their salutes. As they passed Hitler — their once and future countryman — about half of the Austrians gave exaggerated Olympic salutes, while the other half, including all the female Austrians, gave the Nazi salute. The Hungarians and Japanese were also warmly received, but the Czechs and Romanians were not. Not cold but utterly frigid shoulders were reserved for the Australians and the British, clearly nemeses of the burgeoning Third Reich.

Finally it was time for Avery Brundage and his compatriots to march down the red cinder track. Jesse Owens, Ralph Metcalfe, and their fellow sprinters had been excused from the parade in order to rest for their heats the following day, but hundreds of their teammates were smartly dressed in white flannels, blue blazers, and straw boaters. As usual at an opening ceremony, the Americans presented a defiantly unmilitary image. They were barely marching, much less goose-stepping. As they prepared to amble down the track, they were undoubtedly wondering how they would be received. Yes, the Germans had treated them well face-to-face, but it seemed only a matter of time before the residual resentment of the boycott movement would somehow manifest itself. They thought that they would be booed, especially when they passed Hitler.

Since 1908, American Olympic teams had sometimes refused to do what all the other delegations did as a matter of course. When Olympic teams march past the host nation's head of state, their flag-bearers, in a show of respect, dip their flags. But not the Americans—not always, anyway. The American quasi-tradition had been born at the games of the Fourth Olympiad, in London in 1908. That year the American flag-bearer happened to be an enormous Irish-American shot-putter named Ralph Rose. Like most sons of Erin of his time, Rose had no use for the English or their king, Edward VII. When Rose, holding the flag rigidly with one arm, marched past Edward's royal box, his arm and the flag stayed rigid. (He and his teammates had been provoked, it could be argued. Flags of all the participating nations had been raised at the Olympic stadium—all except the Stars and Stripes.) At an Olympic games best remembered for the enmity between the American and British teams, Rose's defiance was the opening salvo (unless we include the flag snub). Explaining Rose's action, or lack thereof, his teammate and fellow

Irish-American Martin Sheridan famously said, "This flag dips to no earthly king." Thus a precedent was set, a precedent that still is invoked by American flag-bearers. A gesture now generally regarded as a token of American independence of thought was in fact mostly a byproduct of Anglo-Irish hostility.

Despite Rose's insistence that the flag dip to no earthly king, it did dip, in 1912, for King Gustav of Sweden (definitely an earthly king), and again in 1924, in Paris, for President Gaston Doumergue (earthly, if not a king). But in 1928, in Amsterdam, the president of the American Olympic Committee banned dipping, insisting on military protocol. Almost eighty years later, Douglas MacArthur's order still stands — although in 1932 the flag was dipped for Franklin Roosevelt, then the governor of New York, at the winter games in Lake Placid, and for Vice President Charles Curtis at the summer games in Los Angeles.

In any case, in 1936, tens of thousands of eyes were focused on thirty-four-year-old Alfred Jochim of Union City, New Jersey. A gymnast, Jochim had won two silver medals four years earlier in Los Angeles, in the vault and the team combined exercise. He looked nothing like Ralph Rose or any of the meaty weight throwers who had so often carried the American flag. He was not quite five foot six. Flanked by Wally O'Connor and Fred Lauer, two burly members of the water polo team, Jochim, strong but slight, struggled slightly under the weight of the flag. As he neared Hitler's tribunal, the spectators wondered whether he would keep the flag upright or not. He did. It did not dip to Hitler. Jochim's teammates, sloppily marching in rows of eight, did snap their heads to the right and, with their right hands, placed their boaters over their hearts in their traditional salute.

Like so many of the best-remembered moments of Berlin's Olympic fortnight, what happened next was remembered differently by different eyewitnesses. In fact, the reaction of the crowd to the American team remains a matter of considerable debate in Olympic circles. According to many observers, the Americans were jeered and whistled at. According to others, they were not.

According to still others, they were greeted unkindly but not vocally. Grantland Rice always stood firmly in the last camp. "It was quite evident that the United States and Great Britain were social outcasts," Rice wrote, "especially the American team, which has been picked to win the Olympics by a big margin . . . [The Americans] received an even fainter demonstration from the crowd than did the British."

Gallico, however, believed that the Americans were simply victims of timing. "There was a moment of silence," he wrote, when the American flag did not dip, "and then was started what I believe would have been a great and genuine roar of welcome, but it was spoiled by an accident of timing and sequence. The German team followed hard on the heels of the Americans. The band quit the march time and burst into 'Deutschland Über Alles' to accompany the German team on its march and the welcome to the Americans was drowned in the greeting to the Fatherland's athletes."

Gallico was outnumbered. Gayle Talbot of the Associated Press, also a witness to the ceremony, wrote extensively about "the debate over whether the American delegation received the German equivalent of the 'Bronx cheer' shortly after passing Chancellor Hitler in the parade of nations. A score of American, British, and German newspapermen as well as spectators were asked their opinion and were about evenly divided on the delicate question," Talbot wrote. One of the German journalists "was positive" that the whistling "was of the barbed variety. 'You must excuse them,' he said, 'because they don't understand America's rule that its flag must not be dipped.'"

With everything else that was going on, Talbot noted that eighty-six-year-old Field Marshal August von Mackensen was "the most uncomfortable looking man in the stadium." Perhaps the finest German commander of World War I, a hero of the Battle of Tannenberg, Von Mackensen was rarely seen without his enormous busby — a relic of his days in the famed Prussian Death's Head Hussar regiment, which he had joined in 1869. The busby, naturally, featured an absurdly prominent death's-head. For some reason, though, Mackensen's busby was conspicuously death's-head-free at the opening ceremony. "Possibly he thought it wouldn't fit into the spirit of the day," Talbot wrote.

Mackensen's legendary stoicism was the perfect counterpoint to Eleanor Holm Jarrett's whimpering. Situating herself near press row, Holm made herself available to any writer who wished to help her wipe away her tears. "I still can't realize that I am out," she told Grantland Rice as she sobbed into his handkerchief. "I never meant anything wrong. I am just as capable of winning here as I did in two other Olympics. I am still ready to win for America and break another record."

"There, there," Granny said, ever the gentleman.

"How could they do this to me? I could start tomorrow and win my race and at least equal the record. Please tell the American public that I was ready and in condition to win honestly. My heart is broken."

"There, there," Granny said. "Don't worry. The Olympic Committee is on the spot, not you."

"That doesn't help me," Holm said, touching his elbow. "They have put me on a cross. I still don't believe it. It just can't be. I did nothing really wrong. I have been foolish but nothing else."

"There, there," Granny said. A few moments later, he described the scene in his column: "The most pathetic figure at the Olympic opening pageantry was Eleanor Holm Jarrett, swimming ace barred by the stupidity of American Olympic officials."

Weeping young women were too much for a southern gentleman such as Grantland Rice to bear. Appalled by the callousness of the AOC and the arrogance of the Nazis, he rather wistfully wrapped up his thoughts on the eve of the first day of competition. "I still believe," he wrote, "the United States will dominate the military-athletic pageant."

When the parade of nations finally ended, a man speaking French bellowed over the loudspeakers. "L'importance aux Jeux

Olympiques n'est pas d'y gagner, mais d'y prendre part, car l'essentiel dans la vie n'est pas tant de conqueror que de bien lutter." These were the words of Baron de Coubertin: "The important thing in the Olympic Games is not winning but taking part. The essential thing in life is not conquering but fighting well." Hitler apparently listened to these words without comment.

After a lengthy, dull address by Lewald, it was the Führer's turn. Mounting the speaker's stand, he said in German, less excitedly than usual, "I announce as opened the games of Berlin, celebrating the Eleventh Olympiad of the modern era." Then, a few moments later, a sleek, handsome, platinum blond, twenty-nine-year-old runner named Fritz Schilgen appeared at the east portal. In his right hand he held aloft the Olympic torch that had been borne all the way from Greece. His white track suit was emblazoned with a German eagle. For a moment, with a light rain falling, he simply stood at the top of the gray steps that led from the portal to the field. "Then," Paul Gallico wrote, "he trotted down the stairs and made a half circle of the track with a pretty stride, the torch dropping little, blazing pieces at his heels."

Nearly 5000 officials and athletes occupied the infield. They stood there watching as Schilgen neared the west portal. Their eyes followed him as he climbed the stone staircase that led to the platform on which the stone Olympic chalice rested on three legs. "For one magnificently dramatic moment," Gallico continued, "he paused there, as true a symbol as ever appeared in human guise, with his torch held up to the gray heavens." Then Schilgen turned toward the enormous tripod and dipped his torch into the caldron. Exploding in blue and orange, the fire took but a moment to catch. Extinguished four years earlier in Los Angeles, the Olympic flame had been rekindled in Hitler's capital.

Wrapping itself in the ceremony, Nazism had enjoyed one of its finest hours. With the assembled multitudes whipped to a frenzy by martial and patriotic pageantry, the scene resembled nothing so much as one of the party rallies in Nuremberg, which had been the goal. While adhering to the letter of the Olympic laws, the Germans nevertheless violated the spirit of those laws by showcasing their Führer above and beyond the Olympians themselves. Clearly, Hitler had been the star of their magnificent spectacle. As Al Laney of the *Herald Tribune* saw it, the ceremony was a wholly successful "demonstration of Nazi organizing efficiency, a personal tribute to Adolf Hitler and pageant such as the modern world seldom has witnessed."

As the crowd inside and outside the stadium drank deeply of the Olympic spirit, Larry Snyder was concerned with a more mundane but nevertheless critical issue. Shoes. Kangaroo-leather running shoes, to be precise. Jesse Owens had no shoes in which to run. That is, he had one pair, which he had been wearing since the Olympic trials, when the other two pairs he had taken east with him were stolen by memorabilia hounds. In its wisdom and parsimony, the American Olympic Committee decided at one of its interminable meetings on the *Manhattan* that the athletes would be responsible for furnishing their own footwear. Sensibly, at another meeting, in Berlin, the committee reconsidered its earlier stance and decided to order one pair of kangaroo-leather running shoes from England for Owens to wear as he attempted to win three gold medals. "If he had to have two pairs," Snyder later recalled, "Ohio State would have to pay for the other pair."

But the English shoes had not arrived, and Snyder was frantically scouring Berlin's sporting goods stores for exactly the right kind of shoe. Finally he located a pair that was just right. Thin kangaroo. Virtually no sole. Of course, there would be no time for Owens to break them in and corns would sprout almost immediately, but Owens did not care all that much. In this respect, he was still very much the sharecropper's son. "Jesse wasn't half so worried about the corns as I was," Snyder said.

By the time Snyder reached the athletes' village, shoes in hand, the opening ceremony was nearly over. He found Owens and Albritton together, in a crowd outdoors, near their brick-

and-stucco cottage, standing in front of what appeared to be a small movie screen.

"What's going on here?" Snyder said. He was disheveled after his long search for the shoes. "What is this thing?"

"I don't know what they call it," Albritton said, "but we've been watching the opening ceremony."

"What?"

"Yeah, coach," Owens said. "It's like radio, with pictures."

"Not very good pictures," Snyder said. He could barely make out anything on the screen, which measured less than a foot across.

They were watching one of the first major television broadcasts ever. The Germans had rolled out their version of the technology just in time for the opening ceremony, with the feed routed via closed circuit to screens in a few select viewing areas. Judging by the image of Hitler that Snyder was trying to see, the fine points of color and tint had not been mastered.

"Is that Hitler?" he asked.

"Guess so," Owens replied. "These the shoes?"

"Of course. I told you we'd find them."

"Where are mine, coach?" Albritton said.

"Yours are just fine. Now, Jesse, be careful with these. Stretch them out good. No blisters. No corns."

Turning his back on Hitler, Owens smiled his winner's smile. "The corns will make me jump farther when they begin to hurt," he said.

15

Day One

THE NIGHT BEFORE he would finally compete at the Olympics, Jesse Owens sat on his bed in his cottage in the Olympic village and chatted with Henry McLemore, a sportswriter who would eventually make his mark as a general columnist for the Hearst syndicate. As Owens spoke, McLemore had to strain to hear him over the noise emanating from the room across the hall. Frank Wykoff had traveled with his phonograph, and he was playing his blues records as loud as possible.

"What was that, Jesse?" McLemore said, cupping his hand behind his ear.

"I said, I always thought I had run as fast as I could," Owens said, raising his voice, "until last Friday night, when I got to Berlin. When we stepped off the train in this foreign country, I knew I could run faster."

"What about the broad jump? Have you done all you can?" It was suddenly quiet. Wykoff was changing records.

"No, I don't think so. Larry is always afraid that I will pull a muscle jumping, so he warned me to be careful and take it easy. In the Olympics, he can't blame me for doing my best, so I plan to let fly all I got."

Scribbling on his notepad, grateful that he had been able to speak to Owens one-on-one, McLemore asked, "What about

Larry? What kind of advice does he give you about your jumping?"

"To me there is no such thing as jumping form," Owens said. "If you can jump, you can jump, that's all there is to it. All I can do is just run as fast as I possibly can, yank my legs up under me at the takeoff, and try to sail as long as I can."

Like so many other observers who had been schooled in the legend of Ferry Field, McLemore subscribed to the theory that an injured Owens was a dangerous Owens. "Maybe it would be a good idea," he said, "to drop you off the roof of this cottage."

"No, sir," Owens replied. "I am going to be in good shape this week. All these years of hard work—I'm going to make sure they were worthwhile."

With that, Owens politely concluded the interview, told Wykoff to knock off the racket, climbed into bed, and quickly went to sleep.

The next morning, as a cold rain fell on Berlin, Owens was up early, eating a breakfast of eggs, bacon, and toast. He drank two cups of coffee. Soon he was on a bus headed to the Olympic stadium. The ride would take nearly forty-five minutes, and he was anxious to feel the Olympic track under his feet for the first time. As the bus rumbled through the western part of the city—really, through its leafy outskirts—he thought about all that was at stake. But he felt no fear. A supreme confidence had washed over him, and the possibility of failure seemed as remote as the cotton fields of Oakville.

Dave Albritton, seated next to him, was jumpy—this would be the only day he would compete at the games, and he knew that it would require a supreme effort to defeat Cornelius Johnson. As he drummed his fingers anxiously on the back of the seat in front of him, Owens was in a reverie. His utter lack of nervousness unnerved his friend.

"Jesse, why are you so calm?" Albritton said. "My stomach is in knots."

"Come on, Dave," Owens said, stretching his shoulders gently, "we just have to do what we always do. The competition was tougher at the trials. There's nobody here tougher than our teammates."

"I guess so," Albritton said as he resumed his tapping. "Still . . ."

For Owens, it would be a busy day but not too challenging. First, at about noon, he would run in the twelfth and final heat of the first round of the 100-meter competition. To reach the semifinals and then the final, both scheduled for the next afternoon, he would have to survive the first round and then the quarterfinals later in the day.

When Owens and Albritton finally emerged from the athletes' bus, they looked up and found themselves unimpressed. Hitler was right. March's stadium was not in the least awe-inspiring. Speer's attempt to dress it up with massive stone panels had been clumsy. More than anything else, it looked like a compromise between architects from two different schools. As they were still taking it all in—the twin clock towers at the east portal, the bell tower to the west, the gargantuan, stylized statues of heroic German athletes on either side of the main entrance—Snyder sneaked up on them.

"Good morning, gentlemen," he said. He was uncharacteristically chipper. Like all the Berliners, he had caught Olympics fever. In fact, he had been at the stadium for more than an hour, trying to determine where Owens and Albritton might rest in the long intervals between the different qualifying rounds, where they would be massaged, where they would warm up.

"Jesse, how do you feel?" Snyder asked, all but ignoring Albritton. He was in his standard summer outfit—white dress shirt, white slacks, dark tie.

"Never better, coach—ready to go," Owens said. His blue sweatshirt was emblazoned with block-faced red letters spelling out *USA*.

"Dave, what about you?"

"Oh, you know, coach, I'm good, just a little antsy."

As Snyder, Owens, and Albritton retreated under the stands, Adolf Hitler was nearing the stadium from the east. He would spend most of the day watching eagerly from his box 50 feet above the field. His distaste for the games had evaporated as they approached. Among the Nazi dignitaries seated with him on that warm, mostly sunny Sunday was Julius Streicher, the virulently anti-Semitic publisher of *Der Stürmer*. Streicher and the rest of the crowd of about 100,000 heiled Hitler upon his arrival and throughout the day. (In fact, Streicher's last words when he was executed in 1946 for war crimes were "Heil Hitler.")

Hitler and Streicher, along with Goebbels and Göring and the rest of the cream of the Third Reich, paid close attention as Albritton, Johnson, and another American, Delos Thurber, took their turns in the qualifying rounds of the high jump, but what really piqued their interest was the shot put, in which a burly Berlin policeman named Hans Woellke was among the favorites. In fact, the Chancellor's eyes were trained on Woellke when, at about noon, the stadium announcer introduced the men who would be competing in the final 100-meter heat. There was Kichizo Sasaki, from Japan, and Dieudonné Devrint, from Belgium, and José de Almeida, from Brazil, and even a sprinter from Malta, Austin Torreggiani. Their names were greeted by the crowd with faint enthusiasm, but when the announcer said, "Yaycee Ohvens, Ooh-Ess-Ah," an enormous cheer rose up, engulfing the stadium. For weeks Snyder had prepared Owens for an icy reception from the Aryan masses, but he had totally misread them. From the first moment they laid eyes on him, they embraced the man who they had heard was the world's greatest athlete.

Absorbing the adulation, Owens turned and looked around the vast arena. Despite all his confidence and calmness, a chill tingled his spine. This, he had not expected. Nor had Hitler, who was more than slightly discomfited by the crowd's reaction to America's black star.

Then Owens walked to the starting line. He looked at the

other men in the field — they were nobodies. There had been better runners, much better runners, at the Big Ten meets. Still, these were the Olympics, so he was methodical. He stretched, he hopped up and down on the red clay, he twisted the knots out of his neck. Then he kneeled into his start position, his heart pumping quietly, his pulse steady, his mind unfazed by the situation. At the gun he was off, an atypically perfect start. Thirty yards down the track, Owens could sense that no one was anywhere near him, but he did not let up — not until the final 20 meters, anyway. Embarrassing the rest of the field, he won by 7 yards — and again the stadium erupted. Most impressive: Owens had equaled his world record — 10.3 seconds — despite the fact that he was essentially running alone, with no one to pace him, and had slowed down near the finish. Clearly there was no problem with the new shoes.

Talking to reporters in the interval between the first round and the quarterfinals, Larry Snyder was asked about the crowd's stunning reaction to the sight of Owens, a man who, according to Nazi ideology, was not quite a man.

"I had braced him for a stony, forbidding silence, because I had read all about the Germanic worship of the Aryan-supremacy idea," Snyder said, keeping one eye on Dave Albritton, who, having qualified for the afternoon finals, was cooling down in the Americans' tiny training room. "But they crossed me up."

After his heat, Owens caught the bus back to the athletes' village, barely, to have lunch and rest before the afternoon quarterfinals. But by the time he got there, it was almost time to leave again. He decided that that was the last time he would leave the stadium after the morning qualifiers. He would just have to tough out the long days at the Reichssportsfeld, as the entire Olympic complex was called.

When he returned to the stadium, he warmed up again in the muggy afternoon air. At four o'clock it was time for his race, and this time he found his rabbit, a Swiss rabbit by the name of Paul Hänni. A moment after the gun went off, Owens could see in his peripheral vision that Hänni was charging hard in front of him — and not noticeably slowing. Suddenly Owens realized that he was in a real race, even as he knew that the top two finishers would advance. Running slightly scared, he somehow reached within himself to find a gear he had never before located. The sight of him running at breakneck speed sucked the air out of the stadium, and as he broke the tape, 4 yards in front of Hänni, yet another roar rose in appreciation. His time was posted a moment later: 10.2 seconds, a new world record.

"No European crowd had ever seen such a combination of blazing speed and effortless smoothness, like something blown in a gale," Grantland Rice wrote. "You could hear the chorus of gasps as he left all rivals far behind."

As the crowd showered Owens with adulation, as he and Snyder embraced, Leni Riefenstahl shook her head in frustration. She wasn't disappointed that Owens had run so brilliantly; she was disappointed that she had not allotted enough cameras to his quarterfinal and would have to make the race more prominent in her film. "I've got to chuck my manuscript," she said. "I'll need all the 100-meter heats for this cutting. This is totally crazy!" It was becoming apparent to her that any documentary about the games would have to celebrate the remarkable black athletes from the United States. This meant another battle with Goebbels, which she dreaded.

Hans Woellke — who had won the gold medal in the shot put, the first Olympic gold medal ever won by a German man in track and field — was summoned to the Führer's box. Riefenstahl entered the box too, where she made small talk with Hitler and Göring. Soon Hitler was also congratulating Tilly Fleischer, who, shortly after Woellke's victory in the shot put, gave Germany its second track-and-field gold medal of the day (and ever), in the javelin. A pretty blonde who had won the bronze medal in Los Angeles, Fleischer so impressed Hitler that he insisted that they be photographed together. Woellke and Fleischer "have been proclaimed national hero and heroine and were welcomed by

Hitler with impressive fervor," Grantland Rice observed as the stadium erupted in cheers and heils for its champions. "I have never seen such a demonstration anywhere at any time before. The outbreak of national feeling is beyond belief."

After congratulating Woellke and Fleischer, Hitler then went a step further. He invited to his box the three nearly identical blond Finns who had swept the medals in the 10,000 meters. Not coincidentally, the Finns were the most Aryan-looking athletes at the games, with the possible exception of Fritz Schilgen and Luz Long.

The commotion in the Führer's box was duly noted by one of its occupants — Henri de Baillet-Latour, who was already in a dark mood. The IOC president had been displeased by the way the Führer had dominated the opening ceremony; now he was convinced that Hitler would insist on remaining the central figure of the games. For the moment, while the Belgian seethed, he said nothing.

Down on the field, meanwhile, his two heats over, his place in Monday's semifinals secured, Jesse Owens covered himself with a towel. "What's for dinner, Larry?" he said, teasing Snyder, completely at ease with the magnitude of the moment. *Ho-hum,* his body language suggested, *another world record.*

"That doesn't win any championships," he told Royal Brougham of the Universal Service a few minutes after the quarterfinal heat. "I'm already worrying about the finals. How did I feel? Well, if you must know, this is my first Olympics, and I was nervous all over. No, I didn't know I was running that fast. I felt easy after the gun cracked and let myself out in the last fifty."

Brougham, like everyone else who saw Owens run in Berlin and who later watched him in Riefenstahl's elegy to his greatness, was most impressed by the graceful, effortless, perfectly coordinated rhythm of his running. "The picture runner," he called Owens.

"Jesse," Brougham now asked, probing unsuccessfully for something more than Owens's steady diet of clichés, "do you think it's possible that you ran *too* hard today? Maybe you should have saved all that speed."

"I don't think I overexerted myself," Owens countered, running the towel through his hair. "I am used to running four times a day, so tomorrow I hope to be rested. I'm going home to lay around now, eat early and then go to bed. This is a great show. I never saw such a wonderful crowd. The track is good and fast and I'll shoot the works in the final tomorrow."

In the final, he said. Assuming that Brougham reported Owens's words accurately, it is worth noting that he did not mention the semifinals. Why would he have? It was clearer now than ever before that no one could keep pace with him.

As the shadows at the stadium grew longer, Cornelius Johnson was, as expected, dominating the high-jump competition. Although he and Albritton shared the world record, Johnson was considered the stronger competitor. (Budd Schulberg, the novelist and Oscar-winning screenwriter, described Johnson, his classmate at Los Angeles High School, as "a funny-looking kid with a boyishly grave face.") As confident as any athlete of his time, Johnson did not bother to remove his sweat suit until the bar had been placed at 2.00 meters (6 feet, 6¾ inches), a height that only he was able to clear. Albritton, for his part, was not at his best. At about seven P.M., with every other event on the first day's schedule concluded, Johnson won the gold medal at 2.03 meters (6 feet, 8 inches). Albritton was in a fierce struggle for the silver medal with Thurber, one of Cromwell's Trojans, and Kalevi Kotkas, of Finland. To determine the silver and bronze medalists, there was a three-way jump-off. Albritton prevailed, and Thurber took the bronze.

Soon three American flags were being hoisted high above the field, and three times "The Star-Spangled Banner" rang out. But unlike Woellke, Fleischer, and the three Finns, the American high jumpers were denied the honor of a congratulatory audience with Hitler. Even before they mounted the medal stand, the Chancellor had left the stadium — to the rapturous delight of the crowd waiting outside. By ignoring Johnson and Albritton, Hitler gave the unmistakable impression that he was both displeased that black men had triumphed at his games and that he would personally have nothing to do with black Olympians — even as he also chose not to congratulate Thurber, who was white. The vast majority of the writers present reported that Hitler had, at the very least, snubbed Johnson — it was unclear in most minds whether Albritton, as a silver medalist, deserved to be congratulated — and criticized the Führer for having done so.

"It isn't for your correspondent to question the movements of the German leader," Jesse Abramson wrote in the *New York Herald Tribune*, "but the fact remains that his abrupt departure after reviewing virtually the entire afternoon program caused the wagging of many tongues." In the *New York Times*, Arthur Daley wrote, "Five minutes before the United States jumpers moved in for the ceremony of the Olympic triumph, Hitler left his box. Press box interpreters of this step chose to put two and two together and arrive at the figure four. In this they may be correct."

But Paul Gallico arrived at a different figure. He was inclined, again, to give the Germans the benefit of the doubt. Referring to Johnson and Albritton, he wrote, "They just missed being presented to Der Führer by a matter of 10 minutes. He had received all the other victors, but the American boys [Albritton and Thurber] took too long at their jump-off and the boss went home to supper. Some of the correspondents saw a political plot in this but deponent is inclined to think Adolf merely got hungry, an old German custom."

Al Laney of the *Herald Tribune* was also unconvinced that Hitler had deliberately insulted Johnson and Albritton. "Since Hitler had been in the stadium nearly four hours at this time and the day's program was practically over and since the Americans did not come forward for the victory ceremony until after 7 o'clock," Laney wrote, "it is unlikely that the disposition to make an incident of the development was justified, although the fact remains that the Americans alone were not received."

A Nazi spokesman denied that Hitler had left his box to avoid congratulating the black athletes. He said that the Führer had to beat the traffic streaming from the Olympic stadium. Even now, no one can say with certainty why Hitler chose not to wait another ten minutes to congratulate Johnson and Albritton. It could have been entirely inadvertent, but it probably was not. Even assuming that the snub was unintended, the question still lingers: Why did it happen? Was it because Hitler thought it would look bad for him to be seen with blacks? Was it because he was genuinely disappointed that blacks had won? Or was it because he was disgusted by the thought of pressing black flesh?

Whatever Hitler's reasons, De Baillet-Latour was furious. His worst fear had been realized. On the first day of the games, a political scandal had erupted. To his credit, he stood up to the Chancellor and demanded that he no longer congratulate any of the medalists unless he planned to congratulate them all. The man who just a few months earlier had boldly marched his troops into the Rhineland, who in two years would annex Austria and force the Czechs to hand over the Sudetenland, who in three years would launch a war that lasted six years and cost tens of millions of lives, meekly acceded to the count's demand. With goodwill the order of the day, Hitler would not allow himself to be called a poor sport. From that day on, he agreed, no athletes would be summoned to his box.

Whether or not Hitler cared to acknowledge them, the achievements of Johnson, Albritton, and Owens on the first day of the games enabled the amateur eugenicists and anthropologists to start spewing their theories again. With Johnson and Albritton winning medals and Owens establishing a world record, much was immediately written about the unique qualities of the black American athletes. But at least Bill Corum of the *New York Journal* declined to subscribe to the nonsense. "Several scientific or semi-scientific magazine articles have appeared recently explaining the rise of the Negro athlete in the past four or five years," Corum wrote in the hours following Owens's Olympic debut. "But it is doubtful if much mystery attaches to it. For the first time these colored boys have had the advantage of equal opportunity and good coaching." Less progressively, he continued, "They were born with natural grace and rhythm, and there's nothing much more important in athletics than rhythm. It also has been written of them that they are closer to the primitive than their white rivals, and, therefore, more rugged. I doubt that part. You don't see them winning any distance races, do you?"

For his part, Jesse Owens was not inclined to pay any attention to the eugenicists. Their theories were all too familiar to him. He was also profoundly uninterested in the snub controversy. By the time Johnson and Albritton had won their medals, he was back at the Olympic village, eating and relaxing to the strains of Frank Wykoff's records. Hitler's failure to congratulate the black high jumpers eventually became the foundation of the story that Owens spent much of his life recounting, but in the hours after it happened, he was fully focused on the races that would take place the following day — the races that would make him an Olympic champion.

Before he went to bed, though, he turned to Albritton. "You know, Dave, there's nothing wrong with a silver medal," he said, sympathetically but slightly unconvincingly.

"Jesse, I'm fine with it," Albritton said, more convincingly. "But what do you know about second place?"

16

Day Two

M ILLIONS OF American newspaper readers had woken up on August 2 to find in their local newspapers a column by Henry McLemore, who was sometimes called "Stopwatch." In his inimitable style, he handicapped the Olympic races as if they were horseraces. For instance, in the 800 meters, McLemore liked John Woodruff, of the University of Pittsburgh: "Big, black, awkward yearling, whose long striding legs will churn cinders in his backwash." Fortunately, McLemore's reverse anthropomorphism did not apply solely to black runners. He called Glenn Hardin, his pick in the 400-meter hurdles, "the best-looking colt at the meeting." In the flat 400 meters, McLemore picked Archie Williams, also a black American, to win and England's Arthur Brown to place: "[Brown] wears blinkers and [is] a tremendous stretch runner." At 1,500 meters, he put his money on Italy's Luigi Beccali: "Much improved over 1932 form when he won in a canter." And so on.

Predictably, McLemore's surest bet was Jesse Owens — but at 200 meters, not 100. Yes, he picked Owens to win the 100 — like almost everyone else — but he observed correctly that the 200 was the cinch. "His best distance," McLemore wrote, "and others appear in it only for the ride." His best parlay for the meet was Owens to win the 100 and the 200 meters.

There was nothing unique about McLemore's views. Ever since Randall's Island and the demise of Peacock as a rival, the press had characterized Owens not as an Olympic gold medal hopeful but as the inevitable triple champion of the sprints and the broad jump. It was the kind of pressure that might make any competitor nervous; the public had been made to think that anything less than three gold medals would be akin to failure. These were also the kind of plaudits that might make any competitor overconfident; Owens had been described as the swiftest, most graceful, most gifted runner since early man first broke into a trot.

But Owens never lost focus or allowed himself to get tangled in the expectations of the writers. By all accounts, he slept and ate well at the Olympic village. He socialized with his teammates and with foreign athletes. He trained under Snyder's gaze but never strained. He didn't even worry about his new shoes. "Larry, they're just shoes," he said, as if he were preparing for a meet against Northwestern or Illinois.

Snyder, in contrast, was nervous — and suspicious. The USC gang made him anxious, especially after the failed power play on the *Manhattan*. It was embarrassing to him that Dean Cromwell was an official U.S. coach while he was merely a personal coach. But what really bothered him was the chumminess between Cromwell and his sprinting protégés, Frank Wykoff and Foy Draper. Snyder knew his disdain for the runners was irrational, based purely on their association with Cromwell. But together Cromwell, Wykoff, and Draper embodied, in his eyes, the enemy — USC and its system. He knew, or thought he knew, what Cromwell thought of him, what he thought of a man who had made his reputation on the talents of Jesse Owens and David Albritton, who were conspicuously darker-hued than Trojan Olympians. Every time Cromwell condescended to him, he detected the unspoken sentiment: *Anybody can win with coloreds; you're not playing by the rules.*

It was no secret that Owens, Ralph Metcalfe, David Albritton, Cornelius Johnson, Archie Williams, John Woodruff, Fritz Pollard, Jr., and Mack Robinson would not have been welcome en masse on Cromwell's team at USC. Of course, they would not have been welcome on *most* campuses. But most campuses were not the home of track and field's most powerful team—and Dean Cromwell was no progressive. To Snyder, Cromwell and Lawson Robertson represented the old guard and all its foibles. Their schools were both private and rich, nothing like Ohio State. But what bothered Snyder most was the way Robertson and Cromwell treated, or rather, seemed to view, Jesse Owens. There was nothing overt. No slurs. No mistreatment. It was tone. It was a look. It was a sense Snyder perceived that in their eyes Owens's genius was diminished by his blackness.

Snyder was certain that Owens detected it too, but Owens was accustomed to that particular attitude and inured to it. His lifelong refusal to allow bigots to truly bother him was often considered, unfairly, a token of his weakness. Even at the age of twenty-two, Owens knew who he was and what he was, and he could see no good reason to allow himself to become embittered by the ignorance of lesser men.

But Snyder was not inured to the double standard, and he allowed himself to feel the anger Owens refused to feel. Now, in Berlin, he thought that he had to protect Owens, not only against Robertson and Cromwell but against Hitler too. He knew that the most powerful people in the stadium wanted to see Owens fail, even as they understood that that was all but impossible. Still, Snyder felt that he had to be vigilant, as if in an instant the Nazis, or Brundage, or Cromwell, or Robertson, might sabotage Owens's Olympic hopes—and his own.

It was only the second day of the games, but already the storylines were developing. They were all about race. America wasn't winning all those medals; its blacks were. This was a fact, but it was also a position. "The American Olympic Committee is still looking for a white man who can even look something like an Olympic athlete," Grantland Rice informed his millions of readers in a column dated August 3. "It may have to call on Scotland

Yard or the German Secret Service. The United States would be below Haiti or China without the darktown parade."

Looking at the bigger picture, Westbrook Pegler was typically incisive. "As events have turned out," he wrote,

> it would have been a great mistake for the Americans to withdraw from the Berlin Olympic Games, for they have had the luck to figure in certain incidents which have redounded to the honor of democracy and the shame of the dictatorial concept of sport . . . The American team will win, thanks to the Negro athletes, whose presence on the squad is proof of the democracy of sports in this country and the result necessarily will discredit dictatorial sportsmanship according to the values established by Adolf Hitler and Mussolini.

Against this backdrop of racial and world politics, Jesse Owens received word that the record he had set the previous day at 100 meters had been disallowed. Officials had determined that the tailwind was above the legal limit. Not a big deal, he said to himself as he loosened up at the Olympic stadium. It was disappointing, yes, especially because the conditions on Sunday had been perfect, and now a morning rain had slowed the red clay of the Olympic track, making record setting unlikely. Unlike Owens, though, the papers, especially the *Daily Worker* and the *Atlanta World,* a black daily, made an issue of the decision, accusing the track officials of bigotry. As usual, Owens chose not to allow himself to get caught up in the carping. He could not afford to. He needed to concentrate all his energies on the events to come.

There were two semifinal heats, beginning at 3:30, and Owens would race in the first against Swedish, Swiss, Dutch, and British runners—and Wykoff. He drew the sixth and outside lane. Wykoff was in the third lane. Franz Miller, the starter, had already endeared himself to Owens with his crisp instructions. Now Miller said, *"Auf die platze"*—on your marks. Then, *"Fertig"*—ready. Then he fired his pistol, and a small puff of white smoke curled into the air. On this day, there was nothing tentative about Owens's start. He anticipated Miller's command perfectly. In white shorts and a sleeveless shirt pinned with his competitor's bib, number 733, all darkened by mud, he hurtled down the track, slowed only by the slippery turf. He overtook Wykoff at the 80-meter mark and finished in 10.4 seconds, one tenth of a second ahead of Wykoff and Lennart Strandberg, the Swede, who also both qualified for the final.

As Owens tried to stay warm in the thick, wet air, he watched the second semifinal. Metcalfe won, in 10.5 seconds, followed by Martin Osendarp of Holland and Erich Borchmeyer, the German Owens had met in 1932.

In less than ninety minutes, at five o'clock, the six qualifiers would meet in the final, perhaps the most glamorous event of the Olympics. For Owens, this would be the first true test of the games. If for some reason he later failed to win the broad jump and the 200 meters, winning the 100 meters would still assure him a place in the Olympic pantheon. Along with the marathon and the 1500 meters, the 100 meters has always been a signature event in the Olympic track-and-field program.

In the training room under the stands, Snyder made Owens lie down on a cot that he had commandeered from a first aid station. With Metcalfe lying a few feet away on the only other cot — borrowed, as it were, from the Spanish delegation, which had returned home when the first shots of civil war were fired — Snyder offered Owens his version of a pep talk.

"Jesse," he said, clasping his straw boater in his right hand, "I want you to focus on your start. You've been doing great, but you have to keep it up. Ralph's running too fast. You can't afford not to be off at the gun. You can't be thinking of anything else. Not the shoes. Not the crowd. Not the girls in the stands. Focus on the gun."

"Coach," Owens said, an impish grin widening across his face, "don't worry. I feel good, as good as I've ever felt."

It was true. He *had* never felt as good. This was not going to be like Ann Arbor — not physically, anyway. For better or worse, he would not be able to channel pain into concentration. His back was fine. He was not distracted. Running on the track back at the athletes' village, he had felt almost weightless, as if gravity were no longer exercising its grasp on his arms and legs and torso. He had achieved a rare state of athletic grace. Spooky, he thought, how effortlessly he had churned down the track and into the air at the broad-jump pit. It seemed to him that he felt like one of Charles Riley's thoroughbreds. Strong. Focused. High-strung, perhaps, but only deep inside, only in a way that would make him go even faster.

Finally it was time to head back to the field. Among the crowds freely milling around the infield was the newest member of the international press corps: Eleanor Holm Jarrett. The International News Service had, upon her fall from grace, hired her as an Olympics columnist, although Paul Gallico and a few other friends were writing all her copy. Like most men, Owens was highly susceptible to her charms, and even as he was preparing for the most important event of his life, he entertained her innocuous questions. Responding to one, he said, "If I get pressed at all, Eleanor, I'll bust that world record all right."

Responding to another — "How *did* you get to be so fast, Jesse?" — Owens said, "The coach at my old high school told me, 'Jesse, you just make believe that track's a red-hot stove lid and let your feet touch it as little as possible. Each foot of that pair you've got should try to beat the other one to it,' so I'm using my feet and legs with one idea, and that is to get there."

Going through his final paces before heading to the vicinity of the starting line, Owens exchanged pleasantries with the other finalists. A fierce competitor herself, Holm watched in awe and admiration as Owens took the proceedings so nonchalantly. "Jesse gave me the day's laugh when he shook hands with another runner before starting," she wrote. "It strikes me he must have been saying goodbye before departing on his fast journey. Hitler and his visiting guests saw the handshake and they had grins on their faces for an hour afterwards."

Ruminating on the enormous crowds gathered at the stadium, Holm reported, "Hitler can sure pack them in—even thicker than Clark Gable." She may have been the only person ever to compare the Führer with Rhett Butler.

Although Owens was clearly the favorite in the finals, a bad start would mean victory for Ralph Metcalfe. It was really that simple. Against Metcalfe, there wouldn't be enough time for Owens to make up for a poor start—maybe against Osendarp or Borchmeyer, but not against Metcalfe. Owens's thoughts drifted back to the previous summer, to the long mornings and afternoons he had spent retooling his start with Charles Riley, who had not been able to travel to Berlin. He thought about all that Riley had done for him and somehow felt his old coach's presence. He looked around and saw, at the edge of the infield, Snyder, who, as always, was worrying enough for them both but trying not to show it. Even from a distance he looked as if he was about to be sick.

For Owens, as for other runners, a perfect start was achieved, technically, only by cheating, by beating the starter's pistol by a split second, undetectable to the human eye. In the heats, Franz Miller had made it easy for Owens to *just* beat the gun. Owens felt it would happen again.

The crowd was in a joyous mood. Karl Hein had just thrown the 16-pound hammer more than 180 feet, establishing a new Olympic record and handing Germany its third gold medal of the games, in the process outdueling his compatriot Erwin Blask.

For Leni Riefenstahl, however, the hammer-throw competition had been ruined when a field judge kicked her cameraman, Guzzi Lantschner, off the field. "You bastard!" Riefenstahl screamed at the judge. "What do you think you're doing?" He did not respond, but he did report Riefenstahl to his superiors.

"It wasn't long before I was handed a note summoning me to the rostrum," she later recalled, "and I feared the worst."

Before she could reach the rostrum, Goebbels cornered her

in the corridor outside. "Just who do you think you are?" he shouted, all but striking her. "Have you taken leave of your senses? I forbid you ever to enter the stadium again! Your behavior has been scandalous!"

Riefenstahl was seething. The subtext of Goebbels's hostility was still her refusal to sleep with him.

"We were given permission," she said, trying desperately not to break into tears. "The German judge had no right to pull the cameraman from his place."

Goebbels was unmoved. "I order you to discontinue your filming here immediately," he said, and then he turned and walked away.

The tears came, and Riefenstahl slumped to the ground. She heard the sound of footsteps. When she looked up, Goebbels was standing before her again.

"Stop crying," he said. "There's going to be an international scandal. I order you to apologize to the judge immediately."

Riefenstahl grudgingly complied. She was eager to get back to work, for she knew, as did everyone else, that the main event of the afternoon was at hand. Finally the 100-meter final was about to begin.

Back on the field, Owens did not seem even slightly apprehensive. Only his muscles seemed to tense, and only when he assumed the set position. His shirt — red, white, and blue stripes running diagonally from his right shoulder to his waist, an Olympic crest over the left side of his chest — was flecked with mud.

With her director's eye for detail, Riefenstahl captured the moment: "The stadium was deathly still. A hundred thousand people were almost afraid to breathe. Metcalfe crossed himself before kneeling down at the starting line; Jesse Owens had the far inside track [the least desirable]. In his white coat, Miller, the starter, gazed in unshakable calm at the runners kneeling in their starting holes."

Then they were off.

Despite the poor conditions, despite his position on the extreme inside lane, the muddiest part of the track, despite the pressure, Owens started perfectly, maybe a nanosecond before Miller pulled the trigger, and instantly surged to the lead. All those hours with Riley the previous summer had paid off. As he climbed out of his crouch and came to his full height, his legs started churning, his stride lengthened, his arms pumped rhythmically, and at 50 meters it seemed to be over. Wykoff was his closest pursuer, a full yard behind. Only by falling could Owens lose—or by turning behind to see the rest of the field. But Owens did not fall. Or look backward. In the final 20 meters, Metcalfe closed the gap slightly, but Owens maintained his form and his focus. With one last surge, he suddenly felt the tape break against his chest. It was over. He had done it. The gold was his. Metcalfe had lost by 4 feet—then had come Osendarp, Wykoff, and Borchmeyer. At 10.3 seconds, Owens's time equaled the world record, a stunning feat considering the muck on which he had run.

Snyder threw his hat into the air and whooped in pleasure. From the crowd, a deep roar rose up and filled the stadium, crashing onto the infield, where Owens soaked it up—a black American in a sea of Aryans. A representative of a supposedly inferior race had won the games' most prestigious competition —and the master race did not seem to mind very much. For the Führer, whose hopes of a Nordic upset had been dashed, the crowd's enthusiasm was probably as confusing as the affection showered on the French in the opening ceremony.

After a moment, Jesse Owens talked to some reporters. "Metcalfe ran a great race," he said. "So did Wykoff." Then, catching his breath, he continued, "This is the happiest day in my life. I guess it's the happiest I will ever have."

Then he slipped on sweatpants and a sweatshirt and moved to a wooden platform where a microphone and a camera had been set up to record his victory statement. "I'm very glad to have won the one-hundred meters at the Olympic games here in Berlin," he said, "a very beautiful place and a very beautiful setting. The competition was grand and we're very glad to have come out on top. Thank you very kindly."

The political implications of Owens's first victory were immediately apparent. "I can't help wondering," Bill Corum wrote from the press box shortly after Owens broke the tape, "if Herr Hitler was thinking about the racial superiority of pure Aryan strains as he saw the Midnight Express whip past. And Jesse, he doesn't even stop to whistle for the crossings."

As Owens mounted the medal stand and the announcer stated that the gold medal had been won by "Jesse Owens, U.S.A." the crowd, perhaps winded by its earlier effort, offered warm if not overwhelming cheers. "Candor compels me to report there was even a finer cheer when Osendarp, the Hollander, standing on the same platform, was announced as having come in third," Gallico reported.

Then, after the medalists were announced, a band struck up "The Star-Spangled Banner." As two American flags were slowly raised—one for Owens, one for Metcalfe—Owens could be seen wiping moisture from his eyes.

When the music ended, he was led by an Olympic official from the field into the grandstand, and it seemed for a moment that, unlike Cornelius Johnson, he might be introduced to the Führer. "There was considerable excitement in the press box when it looked as though local Jim Crow rules might be off to honor Owens' victory," Gallico wrote, "and in charge of an official, he was steered towards the box of Chancellor Hitler in which was also seated Herr Streicher, Germany's number one hater." The writers stood up from their seats to get a better view of the meeting that they thought was about to take place. "However," Gallico wrote, minutes after sitting back down at his typewriter, "Owens was merely led below the honor box where he smiled and bowed and Herr Hitler gave him a friendly little Nazi salute; the sitting down one with the arm bent."

This was the moment Owens would later recall, then choose not to recall — the birth of the myth of Hitler's snub.

Alan Gould of the Associated Press saw it like this: "Chancellor Hitler exchanged hand waves with Owens, who later said the track seemed heavy and he regretted there will not be another opportunity to regain the world record taken away from him because the wind exceeded four miles per hour."

Owens himself disputed that Hitler had shown him any disrespect and told several reporters that Hitler had indeed waved to him.

William L. Shirer, no friend of the Third Reich, wrote, "Hitler's salute to him after the race pleased Owens. He said: 'It strikes me, he's a good sport. I like his smile.'"

"'Mr. Hitler had to leave the stadium early,'" Louis Effrat of the *New York Times* quoted Owens as saying when he was later asked how it had felt to be snubbed by the Führer, "'but after winning I hurried up to the radio booth. When I passed near the Chancellor he arose, waved his hand at me and I waved back at him. I think the writers showed bad taste in criticizing the man of the hour in Germany.'"

Hitler did not receive Owens, but he did not snub him — at least that's not how Hitler's actions were perceived by eyewitnesses such as Gallico, or by Owens. In fact, the *Washington Post* headline that accompanied Gallico's story proclaimed, "Hitler Waives Jim Crow Law to Extent of Saluting Owens."

"Jesse swears he saw Hitler wave to him," Snyder also said. "Maybe he did and maybe he didn't. I wouldn't know. Maybe Hitler made some sort of gesture in the course of his conversation with a henchman sitting near him."

The black press, not quite unanimously, saw a snub where Owens said he saw nothing. The headline in the *Baltimore Afro-American* was typical: "'Adolf' Snubs U.S. Lads."

The *Cleveland Call and Post* was livid. Under a banner headline on its front page that read "Hitler Snubs Jesse," an unidentified reporter wrote, "Herr Hitler, Nazi dictator of Germany, deliberately snubbed Jesse Owens, the outstanding athlete at the Olympic games ... Hitler went out of his way to snub Jesse Owens after his record-breaking run [actually, it was a record-equaling run] in the 100-meter race." The *Call and Post* provided no further details. It did go on to say, "Much feeling has been worked up against the whole American team because they did not dip their colors when the parade on opening day passed Hitler's box, nor did the Americans give the Nazi salute. Then, too, the present German administration is as much anti-Negro as it is anti-Jewish. Therefore, Hitler is trying to avoid showing Negroes any courtesies that may be interpreted as granting them equality."

Among the black papers, the *Pittsburgh Courier* — which, unlike the others, had two correspondents in Berlin — offered a different take. "It has been demonstrated since Sunday that when Chancellor Adolph Hitler left the Olympic Stadium a few minutes before Cornelius Johnson and David Albritton, champion American high jumpers, were to be presented, no snub to them or colored Americans was intended," the *Courier* reported. "On Monday, Robert L. Vann, editor of the *Courier*, cabled this newspaper from Berlin that the German Chancellor had extended the same courtesy and encouragement to Jesse Owens that he gave the other athletes. Hitler was forced to leave the Stadium when he did and Germany officially explained that he did not intend to slight our athletes." In fact, in its edition dated August 8 — the *Courier* was a weekly — a bold-faced banner headline declared, "Hitler Salutes Jesse Owens."

Then again there was the *Chicago Defender*, which, relying on its "foreign press service," and perhaps its imagination, reported that Owens "has captured everyone in Germany but Hitler, who has very conveniently avoided congratulating Owens."

The snub story was everywhere. The editorialist of the *New York Daily News*, which had agitated for the boycott, wrote, "There has been a cloud over the Berlin Olympics largely because of Der Fuehrer, Hitler. He has declined to shake hands with, or receive, our colored athletes because they do not qualify as Aryans. He has conspicuously refrained from saluting them, but he has made it a point to shake hands with the true-blue Nordics. It probably galls Hitler to see his racial theories, as they relate to physical superiority, conspicuously disproved. As host to the 1936 Olympics, he has displayed very bad manners."

William L. Shirer, an eyewitness, said much the same. "Each time Owens trotted up the track after a smashing win the German spectators stood and gave him thunderous applause," Shirer wrote. "But Hitler, I noticed from my seat in the press box a few feet away from him, turned his back to talk to some cronies. One of these was Baldur von Schirach."

A twenty-nine-year-old German aristocrat with an American mother, Baldur von Schirach was the leader of the Nazi Youth and, in Shirer's words, "a handsome young man of banal mind." In 1946 in Nuremberg, after rounding up and sending to their deaths thousands of Austrian Jews, he was convicted of conspiring to commit crimes against humanity. In 1936, though, he had not yet murdered anyone and suggested to the Führer that he allow himself to be photographed with Owens.

The Führer did not just refuse. He was outraged. "The Americans ought to be ashamed of themselves for letting their medals be won by Negroes," he said to Von Schirach. "I myself would never shake hands with one of them."

17

Day Three

J ESSE OWENS had no time to rest on his laurels. There were too many others to be collected. In fact, Tuesday would be the busiest day of his Olympics — and the most consequential of his life. In the morning he would have to scurry back and forth across the track to compete in both the first round of the broad jump and the first round of the 200 meters. Both events were scheduled to begin at 10:30. Then, assuming all went well, he would be back on the track at 3:30 for the 200-meter quarterfinals and then, at 4:30, for the broad-jump semifinals. Finally, the broad-jump finals would commence at 5:45, by which time, assuming he had not been eliminated, he would have run two sprints and made up to six punishing jumps. It was good to be twenty-two and healthy.

The bad news was the weather. The sun peeked in and out of the clouds through most of the morning, but the temperature never rose to 70. It was humid and there was a constant biting wind. Again the conditions at the stadium would not be conducive to the breaking of world records — particularly not in the broad jump. The air was too heavy.

When he arrived at the stadium, Owens took some time to speak with the American reporters. One of them asked if he thought he might be able to equal Alvin Kraenzlein by winning four gold medals at the games.

"I think maybe I can do it," Owens said. "I feel fine. I think I ought to cop the broad jump today and I will run the race of my life in the two hundred tomorrow. Then, if all goes well, I'm going to run in the relay on Sunday, and if we win it, that will make four of those yellow medals." Never before had Owens mentioned the possibility of running in the relay and winning a fourth gold medal. Of course, if he were to run, someone would have to be removed from the relay — someone, in all likelihood, who would not yet have won a gold medal.

For most of the great sprinters who have excelled at both 100 meters and 200 meters, the 200 meters has been the easier race to win consistently. It is also less nerve-racking. Often a runner can overcome a bad start in 200 meters. Luck is less important than it is at the shorter distance. The 200 meters might also be a better test of innate speed. While the 100-meter champions have always been referred to as the world's fastest humans, the 200-meter champions actually run faster. Truly great sprinters are still building speed at the 100-meter mark; the average velocity of 200-meter champions has almost always been greater than the average velocity of 100-meter champions.

Ironically, while Owens is best remembered for his other victories in Berlin, it was the 200-meter competition that he most thoroughly dominated, beginning on the morning of August 4. He was running in the third of eight heats, against runners from Canada, Germany, Great Britain, the Philippines, and Denmark. Only Lee Orr, the Canadian, was of world-class caliber. After observing the first two heats from the infield with his teammates Bobby Packard and Mack Robinson, who would both run later, Owens went to the starting line for his heat. The sun had come out, and the temperature was 66 degrees. He looked at the four white men and the Asian who were also toeing the line. Then he put his head down and waited for the gun.

With every race, Owens grew to appreciate Franz Miller, the starter, more and more. There was the even, precise way he called the runners to their marks, and the predictability of his pacing. He continued to make it possible for Owens to know exactly when the gun would be fired — which is all that a runner can hope for. Now it went off again, and Owens was again off just at the gun, or just before it. He stretched himself out, accelerating smoothly, constantly, and broke the tape 10 yards in front of Orr, his closest pursuer. Orr, in fact, had run the distance in 21.6 seconds, the fourth-best time of the forty-four competitors, but he still lost the heat by a full half second. Owens's time, 21.1 seconds, was a new Olympic record.

Then it was time for the broad jump.

In the sprints, the only athletes Owens considered serious rivals for the gold medals were his fellow Americans. In the broad jump, however, there was Carl Ludwig "Luz" Long. The twenty-two-year-old from Leipzig was not quite Owens's equal as a broad jumper, or the equal of a healthy Eulace Peacock, but he had been having a magnificent season, destroying the German and European records. For what it was worth, Long was the greatest broad jumper the Old World had ever produced. Still, Owens and Snyder were confident that an easy victory was at hand. After all, no one in the history of the world, old or new, had ever jumped as far as Owens; no one had come close. If Owens jumped 25 feet, 6 inches more than a foot less than his personal best, he was still likely to win.

While some observers, such as Henry McLemore, considered the 200 meters Owens's surest bet, another school of thought was in favor of the broad jump. Arthur Daley, for instance, said that the jump was the event Owens was "most certain of winning." And Alan Gould used precisely the same phrase. But about fifteen minutes after Owens arrived at the stadium to begin warming up, he spotted the man who would provide a serious challenge.

Owens and Snyder had both been curious about Long. Like the rest of the track community, they had been hearing his name for months but had never seen him. Most of the pre-Olympic prognosticators had assumed that Long would win a medal — not a gold medal, but a medal. Even Hitler had been talking about Long, whose blue eyes, blond hair, chiseled features, and athletic physique embodied the Aryan ideal. Just three weeks before the games, Long had jumped 25 feet, 7⅞ inches — more than a foot less than Owens's personal best, but still a European record, still long enough to pose a serious threat if Owens had an off day.

As the competitors milled about, Owens and Snyder kept their eyes peeled for Long. Owens knew him when he saw him. A dashing figure in gray sweatpants and coarse black turtleneck, Long was even blonder than he had imagined. He was taller too, an inch or two taller than Owens.

"That's him," Owens said, stretching his arms above his head.

"Long," Snyder said, eyeing the German. "He sure looks like a Nazi."

"He looks like he's in pretty good shape to me," Owens said. It was not his custom to eye his opponents. But there was something about Long. Something compelling — and vaguely threatening. For one thing, there was his Aryan coloring and features. And as confident and relaxed as Owens was, it worried him to know that Long had the advantage of not having been recently at sea, that he had been eating familiar food and would be competing in front of his own people.

Years later, recalling the moment when he first caught a glimpse of the German, Owens wrote, "Long was one of those rare athletic happenings you come to recognize after years in competition — a perfectly proportioned body, every lithe but powerful cord a celebration of pulsing natural muscle, stunningly compressed and honed by tens of thousands of obvious hours of sweat and determination. He may have been my archenemy, but I had to stand there in awe and just stare at Luz Long for several seconds."

Perhaps not since he had first competed against Ralph Metcalfe had Jesse Owens been in awe of anyone. It was the other guy who was supposed to be in awe of *him*. But Long was not awed. He was poised and self-assured. More than his physical bearing, it was Long's confident mien that truly made Owens uncomfortable. *How can he look so calm?* Owens, renowned for his evenness, thought to himself. *Doesn't he know who he's up against?* Long unnerved him, and he was suddenly and sickeningly reminded of the way he had felt a year earlier when he was running and jumping against, and losing to, Eulace Peacock. Long, like Peacock, was conspicuously bereft of the fear Owens had grown accustomed to recognizing in the eyes of his competitors. But against Peacock, Owens had been worn out; there was no such excuse now. And in an instant, he could actually feel all the confidence he had built up drain from his body. For a week, everything had seemed so easy and natural. His mind and his body had been in perfect sync. Now, like a golfer suddenly afflicted by the yips, he was filled with uncertainty. He tried to banish the negativism from his thoughts, to no avail. *Okay, don't worry,* he said to himself. *This is my best event, I am the best ever, I have jumped almost 27 feet.* For a moment that thought soothed him, and he allowed himself to believe that mentally he was back where he had come from, in that zone where everything was effortless.

Considering that the spectators would see only preliminaries, the crowd that morning was enormous — about 90,000 strong. Still, compared to the finals for the 100 meters — when about 110,000 people had crowded Werner March's bowl — the atmosphere was intimate, and having already claimed one gold medal, Owens should have felt fairly relaxed. Even if the world was expecting three gold medals from him, no matter what happened now, he would still walk away with the most prestigious championship in sports. Additionally, to qualify for the afternoon semifinals, competitors needed to leap only 23 feet, 5½ inches. That was a distance Jesse Owens had been regularly surpassing since high school — as Charles Riley, if he had been in Berlin instead of in Cleveland, would have reminded him; as Larry Snyder, at that moment, did remind him.

"No problem at all, Jesse," Snyder said, employing his gentlest tone, fretting inside—because that's what he did—but outwardly the picture of calm. "Smooth and easy. And remember, don't overdo it. You don't need to set any records this morning—they won't do you any good in the finals." But Snyder could not conceal his concern from Owens, who in turn tried to comfort him.

"Come on, coach," Owens said, "Twenty-three five? I won't strain anything."

Asking Owens to jump such a short distance was akin to asking him to run the 100 meters in 11.0 seconds—a staggeringly easy threshold. When his turn came, Owens, still in his sweat suit, scampered over to the runway. His mind still racing, filled with visions of Luz Long, he did what he always did—what all American broad jumpers always did: he jogged down the runway and through the landing pit, just to feel the path under his feet, just to get a sense of where he was. But as Owens stepped out of the landing pit, he turned and, to his horror, saw the white-jacketed official who monitored the jumps holding aloft a red flag. He was signaling a foul.

"What?" Owens nearly shouted. "What do you mean? Why?"

It was explained to him, in broken English, that his practice "jump" had counted. The American custom of running through the pit was unknown in Europe. Robertson was in the officials' faces, demanding that the jump not count, but his entreaties were dismissed. The rules were the rules—especially in Berlin. Snyder wanted to berate Robertson for not knowing the European rule, but at that moment he couldn't get to him, so he turned his anger inward, berating himself. *Dammit*, he thought, *now Jesse has only two chances to jump the minimum distance required to reach the finals.*

Owens tried to calm down. "Forget it," he said to himself. "Get that out of your mind. Just bear down."

But Owens was upset about the foul and nervous about

Long, who had already qualified easily and for whom the crowd had cheered wildly. It had been a long time since anyone had mounted a serious challenge to him in the broad jump — not since Peacock. "Second by second," he later wrote, "home seemed farther away. Much more than the 6000 miles. I wanted to be here, in Berlin, in the Olympics, but it wasn't my turf. It was Luz Long's turf." A sure thing minutes earlier, the broad jump was now his moment of truth.

For any athlete, self-doubt is deadly. When technique and timing are essential, as they are for long jumpers, self-doubt is doubly lethal. No jumper wants to be thinking as he's charging down the runway. He wants his reflexes, not his brain, to do all the work. Now Owens was thinking too hard — and all his thoughts were negative. His rhythm was off. His strides were choppy. His form was flawed. In his mind, he could suddenly do no right. *Stop thinking those thoughts,* he said silently to himself. This was one of those moments, again, when he reached into his reservoir of Charles Riley bromides. *Don't worry about your opponent,* he thought. *Just do what you are capable of doing.*

Soon it was time for his second attempt. He set himself, then started running, picking up speed, but he couldn't stop those thoughts. As he approached takeoff, he found himself unconsciously measuring his strides, trying to space them so he would hit the board cleanly but not beyond the line. As he kicked into flight, he knew he had not fouled. He also knew that his gait had been too hesitant. It was a wobbly jump, and as soon as his spikes hit the sand he knew he was short. He had jumped only 23 feet, 3 inches — 2 inches short of what he needed, a laughable effort for the world-record holder. Long's qualifying jump had been more than 2 feet longer. "The situation," Arthur Daley wrote, in his understated style, "was getting to be alarming."

Hanging his head, Owens could barely fathom what was happening. *No,* he thought as he kicked at the dirt, *this isn't why I came here! Did I come all this way for this? To foul out of the trials and make a fool of myself?* He tried to talk himself out of his funk, reminding himself that he was the greatest long jumper ever, that no one else in the competition was comparable, but his mind kept returning to the sight of the red flag—and to the sight of Luz Long.

Then Owens felt a tap on his shoulder.

"What has taken your goat, Jazzee Owenz?" the stranger said slowly in a German imitation of British English, his accent thick but understandable. He was wearing a white shirt emblazoned with an eagle and a swastika. "I am Luz Long. I think I know what is wrong with you."

"Hello, Luz," Owens said. At moments like this, even under enormous pressure, he could project infinite calm. This time, Long's casual introduction really did drain all the tension from him.

Matter-of-factly, Long said, "You know, you should be able to qualify with your eyes closed. Why do you not draw a line a few inches in back of the board and aim at making yourself take off from there? You'll be sure not to foul, and you certainly ought to jump far enough to qualify."

"The truth of what he said hit me," Owens later said. "I drew a line a full foot in back of the board."

In some versions of this oft-told story, Long placed Owens's sweatshirt behind the board as a visual aid, but no one who was covering the event mentioned such a gesture in their stories. It seems much more likely that Long simply offered Owens some friendly advice—an act of sportsmanship that embodied the Olympic spirit.

Among the myriad technical innovations Leni Riefenstahl pioneered specifically to chronicle the games, she had had a trench dug near the broad-jump pit so that her cameramen could capture the jumpers in midflight from a low angle. Now her cameraman was frantic. Like everyone else in the stadium, he knew that Owens had only one chance remaining. This was a shot he could not miss. If he did, Riefenstahl would never allow him to forget it. He could already hear her: "Did you get it? Did you get it?"

Seated in the stands, Snyder wished he could tell Owens to forget where he was, just to pretend they were in Columbus — and not to foul. Whatever else, not to foul. To give himself some room near the board. But then he realized that it would be best to give Owens a moment to himself. What he needed was not advice but simply a cocoon in which he could gather himself.

Now Owens was alone — the noise of the crowd blocked, his peripheral vision narrowed to the length of the path at his feet. From his spot high above the field, Grantland Rice tried to locate on Owens's face "some telltale sign of emotion," but found none. Then Owens took off, building speed, measuring his steps, looking for that spot behind the board. Timing his strides perfectly, he leapt from well short of the board and sliced through the muggy air, his legs folded beneath him. An instant later he was crashing into the pit. He knew immediately what had happened. He had succeeded. In fact, he had jumped more than 25 feet, despite allowing himself about a foot of leeway on takeoff. Riefenstahl's cameraman had the shot, but now it probably would not make the cut — just another preliminary jump.

As Snyder breathed deeply and shook his fist, Luz Long went over to pat Jesse Owens on the back. "See," he said, "it was easy."

Owens just smiled and clasped Long's hand in both of his. *"Danke,"* he said. It was the one German word he had picked up. Nearly forty years later, when he wrote his memoirs, he pointed to Long's gesture as the defining moment of his Olympic experience — and his life.

In the meantime, though, Owens still had to spend the long afternoon waiting around the Olympic stadium for his 200-meter quarterfinal race and then the broad-jump semifinals and, he hoped, the final. Disaster averted, he and Snyder and Albritton, who had remained in Berlin mostly to watch Owens, rested in their training room, Owens napping on the Spaniards' cot.

For Snyder, the interval between the morning and afternoon events was a time to contemplate his own failings. He felt that he had let Owens down by not knowing and warning him that he could not make a practice run through the pit. Yes, it had all turned out okay, but still, Owens's scare gnawed at him. He pledged to himself that there would be no subsequent oversights; he would be vigilant.

At 3:45, though, in the third heat of the 200-meter quarterfinals, there was no need for any coaching whatsoever. Owens simply outran his five competitors, again finishing in 21.1 seconds, equaling the Olympic record he had set five hours earlier. Again the crowd embraced him as if he were one of its blond, blue-eyed Teutons. He felt good. The anxiety of the morning had dissipated. Now all that remained before the sun set was the broad jump.

As Owens was setting records on August 4, so too were many of his teammates. John Woodruff, another black man, won gold in the 800 meters, as expected, and Glenn Hardin, who was white, won gold in the 400-meter hurdles.

Most notably, Helen Stephens broke the world record on her way to winning the gold medal in the women's 100-meter dash—defeating Stella Walsh, the 1932 gold medalist, whose family, like Owens's, had moved when she was a child to Cleveland. Unlike the Owenses, the Walasiewiczes were not escaping the prejudices of the Deep South. They were escaping the hopelessness of Poland under the Russian czar. Walsh had been in Cleveland since the age of one but nevertheless competed for Poland. In 1980, after she was shot to death, an autopsy revealed that she was as much male as female. Strangely, in the aftermath of Walsh's Olympic defeat, a Warsaw newspaper ungallantly claimed that *Stephens* was a man. German officials, however, leaped to her defense by announcing that, in the words of *Time* magazine, "they had foreseen the dispute, investigated sprinter Stephens before the race, [and] found her a thoroughgoing female."

Gender uncertainty was actually a major issue in women's athletics at the time. In addition to the whispers, and shouts, about Stephens and Walsh, an elite Czechoslovakian runner who had competed as a woman decided that she would be happier living as a man — "I argued with her but lost the decision," Ted Meredith, her coach, said — and an official statement of the International Olympic Committee regarding the gender of a Japanese broad jumper who had competed in the women's event at the games in Amsterdam in 1928 referred to said competitor as "It."

An eighteen-year-old Missourian known as the Fulton Flash, Stephens was received by Hitler, unofficially, after her record-setting win, "reviving charges Owens and the Negroes are being discriminated against on racial grounds," Davis J. Walsh reported for the International News Service, "although some sources say the Afro-American victors will be received later by the Fuehrer at his convenience."

"Boy, what a thrill," Stephens told reporters after shaking Hitler's hand in a room under the stadium, away from the photographers' lenses and the glare of De Baillet-Latour. The Führer, in his simple brown uniform, was accompanied by Rudolf Hess, his powerful deputy and oldest comrade. (In 1920 Hess had become the sixteenth member of the Nazi party.) Hitler and Hess both saluted Stephens, who had thrown a pair of slacks over her running shorts. The conversation, though, did not go anywhere. Stephens could not speak German, and Hitler and Hess could not speak English. "I said something and I guess he congratulated me," Stephens said. "Anyway, I heard some interpreter say so."

Stephens did not linger long with the Führer. She had to get back to the field, where she was supposed to be throwing the discus. Still winded after running faster than any woman ever, she did not throw it very far.

Like so many American athletes at the games, Stephens had been impressed with the Germans' organizational genius and the reverence in which they held their leader. Naturally, she was thrilled to be singled out for congratulations. "It's enough in any girl's life to break a record," she said, "but getting a handshake and a pat on the back from a big man like Hitler is just about my speed for one day."

Among the other winners that afternoon was Germany's Gisela Mauermayer, who pleased the Führer by capturing the discus championship. "It is something," Rice wrote after witnessing the medal ceremony, "to hear a hundred thousand voices singing 'Deutschland Uber Alles' as the swastika flag catches the wind and a hundred thousand hands extend in the Nazi salute."

It was against this backdrop that Owens and Luz Long would duel in the broad-jump final, which consisted of two rounds. At 4:30, in the round that would reduce the field from sixteen to six, they each jumped three times, Owens surpassing the previous Olympic record with his second jump, leaping 25 feet, 10 inches, and Long surpassing it with his third, jumping 25 feet, 8¾ inches. With each prodigious jump, the cheering of the crowd shook the stadium to its core. Then, at 5:45, less than thirty minutes after their final jumps in the elimination round, it was time for the final round, in which they would each have three more attempts. The German jumped first, clearing 25 feet, 4 inches — not good enough. After Owens faulted on his next attempt, Long jumped again, and this time he too jumped 25 feet, 10 inches — sending the crowd into hysterics. The gold medal would be decided in the final jumps.

Now brimming with confidence, Owens took his turn and again smashed the Olympic record, jumping 26 feet, 0 inches. Down to his last chance, Long, not a particularly fast runner, sped along the track, jumped — and faulted. Just like that, Owens had won the gold medal, his second of the games. But he still had one jump remaining. Instead of forgoing it — the sensible choice — he charged down the runway one more time and leaped as he hit the board. "As he hurled himself through space," Grantland Rice wrote, "the Negro collegian seemed to be jumping clear out of Germany. The American cheering started while Jesse was airborne."

The measurements confirmed what had been clear even to those seated in the farthest reaches of the stadium. Owens had smashed the Olympic record again, jumping 26 feet, 5½ inches. In the space of two hours, he had taken five jumps, all of them better than the previous Olympic record. His record — though 3 inches shorter than his jump in Ann Arbor fourteen months earlier — would stand for twenty-four years. (Ultimately, the marks of Owens and Long would not count as Olympic records because of the tailwind.)

Before Owens could dust himself off, Luz Long ran to congratulate him, throwing his right arm around Owens's shoulders. Then, turning to the side of the stadium where Hitler was seated, Long clutched Jesse's right hand with his left and hoisted their arms into the air. Together, Long and Owens paraded across the infield, hand in hand.

It is impossible to say whether the Chancellor noticed Long's eternal act of sportsmanship. Regardless, he was so impressed by the silver medalist's efforts that he insisted on congratulating him, in the private room behind his viewing stand, before leaving the stadium. Since Owens had won again, the American press corps wanted to see how Hitler would react. "His eagerness to receive the youthful German," Arthur Daley wrote, "was so great that the Führer condescended to wait until his emissaries had pried Long loose from Owens, with whom he was affectionately walking along the track arm and arm. All the Negro received was his second gold medal, which probably satisfied him well enough at that."

Now that it was apparent that Jesse Owens was the true superstar of the games, the Nazi press, which had toned down its rhetoric for several days, turned it back up. In particular, *Der Angriff* (*The Attack*) lived up to its name. "If America didn't have her black auxiliaries, where would she be in the Olympic Games?" *Der Angriff* asked after Owens won the broad jump. Conceding that the Americans were likely to continue to win medals, the paper petulantly pointed out, "But if the Americans hadn't enlisted her black auxiliary forces, it would have been a poor look-out for them. For then the German, Luz Long, would have won the broad jump; the Italian Mario Lanzi, the 800 meters run, and the Hollander Martin Osendarp, the 100 meters. The world would then have described the Yankees a great Olympic disappointment. It must be plainly stated that the Americans aren't the athletic marvels we thought they were despite Owens, Metcalfe, Woodruff and Johnson."

For more than seventy years, *Der Angriff*'s diatribe has been cited as the ultimate example of German hostility to Jesse Owens and his black teammates. But its words, if not its tone, were nearly identical to those of Grantland Rice, whose oft-used phrase "darktown parade" was in its own way as offensive as *Der Angriff*'s "black auxiliaries."

Rice — and, to be fair, most of his colleagues in the American media — was as preoccupied with the racial question as the staff of *Der Angriff*. Describing the atmosphere at the Olympic stadium on August 4, he wrote, "Tuesday was a dark, raw day of rain and wind, but it looked even darker to the fifty other nations participating in the Olympic Games as our Ethiopian troops continued their deadly fire." He wrote that Glenn Hardin, the champion in the 400-meter hurdles, "startled the German multitude by proving that the United States had a white man who could win."

Hardin's victory notwithstanding, Rice lamented the decline of the white American track-and-field star, which must have come as something of a shock, considering how thoroughly white Americans from Alvin Kraenzlein to Johnny Hayes to Ray Ewry to Charley Paddock had dominated the Olympic track-and-field competitions since their revival in 1896. "The collapse of the American whites has been terrific," Rice wrote. "Apparently the race here is to the swift, and the black and sepia are too strong." And then a nod to Kipling: "The white man's burden has broken the white man's back as far as America is concerned; the United States would be outclassed except for our black-skinned frontal and flanking fire."

Joe Williams, Rice's counterpart at the *New York World-Telegram*, was equally blunt. "It begins to look as if we will have to make the 'Darktown Strutters' Ball' the official hymn of the American Olympiad," he wrote. On the subject specifically of Owens, he was more benign, continuing, "I do not like to invite unsavory criticism by discussing the beauty of a man's legs, but no worker in bronze could improve on the gracefully powerful lines of this young gent's underpinnings. He is built like a Man o' War colt and geared just as high for speed."

After the broad-jump medal ceremony, Rice was able to track down Owens for a quick interview. Like most of his colleagues, he was more curious about Owens's perception of his treatment at the hands of Hitler than about his struggle to qualify for the broad-jump semifinals.

"I haven't even thought about it," Owens told Rice. "I suppose Mr. Hitler is much too busy a man to stay out there forever. After all, he'd been there most of the day. Anyway, he did wave in my direction as he left the field and I sort of felt he was waving at me. I didn't bother about it one way or the other."

While Owens continued to deny that there had been a snub, it remained a dominant theme in the papers back home. Itching for a fight, the *Daily Worker* decried the Führer's treatment of Owens, continuing to ignore Owens's own account of what had happened. The *Worker* gave front-page coverage to Angelo Herndon, the chairman of the Youth Committee of the Communist Election Campaign Committee, when he issued a statement simultaneously calling for the defeat of Alf Landon in the 1936 presidential election and condemning Hitler for his "insults" to Owens and other black American athletes in Berlin. "The crowning achievements of Jesse Owens and other American Negro athletes have been ignored by Hitler," Herndon said in his statement. "Leading Nazi papers carry vicious headlines such as 'German is first white man to finish' when announcing results of a contest in which a German [Long] finished second to Owens." Then, because he couldn't help himself, Herndon equated

Landon with Hitler. "The Nazi Olympics," he said, "have given American Negro youth a picture of what they may expect in the United States if the reactionary forces represented by the Liberty League-Hearst-Landon combination triumph." (Apparently Jesse Owens didn't make the connection. He campaigned for Alf Landon, against Franklin D. Roosevelt, after he returned from Europe in late August. In fact, Owens famously said while on the stump, "Hitler didn't snub me — it was our president who snubbed me. The president didn't even send me a telegram.")

Despite the triumphs of Owens, Woodruff, and Johnson, the games were going along smashingly for the Germans. All that they had hoped to achieve they were achieving — and more. Not even the most optimistic among them had expected their athletes to win so many medals of all hues. More important, the Germans were proving — as Jeremiah T. Mahoney and William L. Shirer and so many others had feared — that their new regime was capable of a certain kind of greatness. "The games were overshadowed," Thomas Wolfe wrote, "and were no longer merely sporting competitions to which other nations had sent their chosen teams. They became, day after day, an orderly and overwhelming demonstration in which the whole of Germany had been schooled and disciplined. It was as if the games had been chosen as a symbol of the new collective might, a means of showing to the world in concrete terms what this new power had come to be."

But at the center of it all, Wolfe wrote, was Jesse Owens: "Everywhere the air was filled with a single voice. The green trees along Kurfurstendamm began to talk: from loudspeakers concealed in their branches an announcer in the stadium spoke to the whole city . . . *Owens — Oo Ess Ah!*"

Yet the power and prestige of the Third Reich were far from Jesse Owens's thoughts as he retreated to the athletes' village after a grueling and magnificent day at the stadium. As tired as he was, he made a point of seeking out Luz Long. He found him in his cottage, reading on his bed. Long got up, embraced Owens, and in his halting English invited him to sit down. For the next several hours, Long and Owens communicated as best they could in English, sharing their hopes and their fears, opening up to each other in ways they had rarely, if ever, opened up to anyone else. One was American, the other German, one black, the other white, one poor, one middle-class, but they found common ground as athletes and sportsmen in a world increasingly dominated by dictators. In the face of so much global uncertainty, they could sense that their Olympic medals would do little to shield them. Owens, after all, was going home to a country deep in depression, in which his skin made him an outcast. More politically aware than Owens, Long was deeply conscious of what his country was becoming and afraid of where it was going. His father's generation had been all but wiped out in the trenches, and yet less than twenty years later, his most powerful countrymen were in another bellicose mood.

Regardless of the distance and differences that divided them, Owens and Long made a pact to remain in contact. The adrenaline coursing through their veins had long since been drained. Now they were simply exhausted. They said their goodbyes, and Owens went back to his cottage. As he tried to fall asleep, though, he could not keep his mind from replaying all the day's dramas. There was Hitler and Long, the races and the jumps. Eventually, finally, his brain shut down, and he slept.

18

"He Flies Like the
Hindenburg": Day Four

AFTER THE HIGH DRAMA of the 100-meter finals and the broad-jump competition, even Owens expected the 200-meter semifinals and final to be anticlimactic. Since he did not have to be at the stadium until 3 P.M., he tried to sleep in — but it was impossible. His cottage had become a landmark in the Olympic village. Fans and fellow athletes were constantly peering in his windows, trying to catch a glimpse of the world's greatest athlete. When the commotion eventually woke him up, he did so to the thought that in all likelihood this would be his final day as an Olympic competitor; his life's work was within hours of completion. It seemed inconceivable that he would fail to win his third gold medal. After all, there were only two men in the world he considered capable of defeating him at 200 meters, and neither would be in the race. Eulace Peacock was home in New Jersey nursing his torn hamstring, and Ralph Metcalfe would be a mere spectator, having failed to qualify for the 200 meters back on Randall's Island. Bobby Packard and Mack Robinson were decent enough runners, but they inspired in Owens no dread whatsoever.

In this event, if in no other, Owens would be competing for records more than medals. But again it seemed the weather — the coldest ever at the summer Olympics, and pouring rain — would not cooperate.

Meanwhile, back at home, Owens's victories at 100 meters and in the broad jump had made him an icon overnight. "It is my pleasure to convey to you the congratulations of the people of your State for your brilliant achievements in the Olympic Games," Governor Martin L. Davey of Ohio cabled to Owens. (There was no similar telegram from the governor of Owens's native state.)

In Washington, Shirley Povich of the *Post* was among the many who had decided to cast Owens's victory as nothing less than the triumph of good over evil. "Hitler declared Aryan supremacy by decree," Povich wrote, "but Jesse Owens is proving him a liar by degrees." Povich went a step further, lauding Owens at the expense of the man he had suddenly replaced as the world's preeminent black athlete.

> It was a year ago at this time that Washington Negroes couldn't see Jesse Owens in their midst for craning their necks at Joe Louis. It struck this department at the time that the colored folks of the Capital were shamefully neglecting a man of their race who was destined for even greater fame than Louis. But a year later finds their positions reversed. Jesse Owens, making Olympic history, with two titles already won and another looming today, is quite the undisputed idol of his race, or should be.

Back in Berlin, American reporters asked Louis's vanquisher what he thought about all those black men winning gold medals. "They are great and Mr. Owens is the most perfect athlete I have ever seen," Max Schmeling replied. Then, searching for exactly the right simile — albeit one that in nine months would be outdated by tragedy in the sky over Lakehurst, New Jersey — the Black Uhlan added, "He flies like the *Hindenburg.*"

Of course, Owens's victories were only part of the biggest

story of the games—the unprecedented brilliance of the so-called black auxiliaries. As Robert L. Vann declared in the *Pittsburgh Courier*, "America's 'athletes of bronze' are marching today—AND THEY CAN'T BE STOPPED!"

In *The New Yorker*, Janet Flanner, writing under the nom de plume Genet, was characteristically droll as she delighted in tweaking the Nazis. "Though it can't be what Germany arranged the games for," she wrote,

> the racial superiority of the Negro athletes has so far been the signal ethnological demonstration of Berlin's Olympiad. Owens, Johnson, Metcalfe, Albritton, Woodruff, as alike in their stylized physique as the dark archaic figures on an Attic vase, have established in the sporting arena a sort of new muscular mythology in which they are the fast and far-leaping gods, against whom pale mortals haven't a chance. In their events, Negroes have given not performance but phenomena.

Even as Owens was adding to his collection of gold medals, he too was putting his achievements in the larger context. In an open letter to the *Courier*'s readers, he wrote, apparently with Vann's assistance, "I am proud that I am an American. I see the sun breaking through the clouds when I realize that millions of Americans will recognize now that what I and the boys of my race are trying to do is attempted for the glory of our country and our countrymen. Maybe more people will now realize that the Negro is trying to do his full part as an American citizen."

Here again was an example of Owens's careful politics—the tone one of conciliation rather than confrontation. Although this tone would diminish him in the eyes of the black militants of the 1960s, in Berlin in 1936, Owens was a true revolutionary, fighting against the ugliest regime on the planet, embarrassing Hitler, Goebbels, Streicher, and the rest of the Nazi leadership simply by being at his best.

None of this had weighed on Owens in the first few days of competition. He had been content to win. Now, though, he was the leading figure of the games, the man who was standing in for all minorities everywhere in their struggles against tyrants. He had gone to Berlin expecting to compete against Metcalfe and Long; finally he was willing to accept that all along he had been competing against Hitler. Long's skepticism about the regime, expressed in their late-night conversation, had hardened Owens's attitude. He did not quite know where his new feelings would take him or how he might act on them, but he knew suddenly, at some level, that he had been politicized. Snyder had always downplayed the politics of sport—not unlike Brundage—and Owens had been happy to be guided by self-interest. With his medals in hand, though, he was finally secure enough to express his real opinions.

At noon he caught a bus to the stadium and spent most of the ride posing for pictures and signing autographs. If anything, his fellow Olympians were more awed by him than the general public was. "It seems to me that Jesse is spending most of his time here smiling for the birdie," Snyder carped to a reporter. Owens, through it all, kept smiling.

"Hey, Jesse," Snyder said when they found each other at the stadium. "How'd you sleep?"

"Great, coach," Owens lied. "I feel terrific."

"You don't look it."

In the semifinals, Owens looked good but not quite terrific. Surprisingly, he did not even equal the Olympic record he had set the day before, running the curved sprint in 21.3 seconds, two tenths of a second slower than the previous day. Meanwhile, in the other semifinal, Mack Robinson tied the record and Canada's Lee Orr ran, like Owens, a 21.3. It occurred to Owens that winning his third gold medal might not be quite so easy. But that thought was fleeting. He would have almost three full hours in which to nap before the final, which he assumed would be his last Olympic event. With a little shuteye, he would be back on the track at his best for the final.

To be sure, Mack Robinson was not the equal of Metcalfe or

Peacock, but he was a prodigious talent nevertheless. Despite a congenital heart defect, he was among the world's fastest men, and he would eventually jump more than 25 feet. A mere freshman at Pasadena Junior College, he had had to find two local businessmen to finance his trip to Randall's Island and even in Berlin often seemed to be the forgotten man among the American sprinters. As Robertson, Cromwell, and Snyder were fighting over Owens on the voyage to Europe and in their first days in Berlin, Robinson had felt largely ignored by the cadre of American coaches — and later said so.

But Robinson needed more than a coach could offer to defeat Owens that evening at the Olympic stadium. Owens ran arguably his greatest race of the games. On a muddy track, heading into the wind, with the temperature dipping to 55 degrees, he ran a blistering 20.7 seconds, shattering the world record for the distance around a curve. In fact, his time was only one tenth of a second slower than the record for 200 meters on a straightaway. Robinson also ran brilliantly, but his 21.1 put him well behind Owens at the tape.

"[Owens] looked like a dark streak of lightning," Rice wrote. "His final performance in breaking all records for 200 meters under miserable conditions, with a rain-soaked track, left the athletes of fifty-one nations goggle-eyed with astonishment."

As Owens crossed the finish line, the stadium again erupted in appreciation of his genius. Tens of thousands of Germans rose to their feet, whistling because they thought it was what Americans did, cheering, as best they could, a black American. Some reporters took German support for Owens as a sign that the government's discriminatory policies did not reflect the will of the people. "If the Olympics clearly demonstrate this," Frederick T. Birchall wrote in the *New York Times,* "and this perhaps brings about some amelioration of the racial persecution in Germany, the location of the games in Berlin will have had its advantages. However, this is almost too much to hope for, because after the tumult and the shouting of the Olympics is over, all the realities of German politics will come to the fore again." Birchall could not have been more prescient.

Among the non-Germans moved by the sight of Owens's crowning triumph was Thomas Wolfe. Seated in the box of William E. Dodd, the U.S. ambassador to Germany, Wolfe howled in delight, so loudly that the ambassador's daughter, Martha (later allegedly a Soviet spy), said that Hitler heard him and turned around in anger. "Owens was black as tar," Wolfe wrote, "but what the hell, it was our team, and I thought he was wonderful. I was proud of him, so I yelled."

Owens had done it — three events, three gold medals. He had achieved all that he had set out to achieve. Before Snyder rushed to congratulate his magnificent charge, he bowed his head and clenched his fist. His mission too was accomplished. Owens had won three individual events at a single games. The writers had already nearly exhausted their arsenal of verbal laurels — but more were located. Alan Gould got it right: "Incomparable," he wrote. "Matchless."

Still, Owens all but apologized for not breaking the world record in the 200 meters. "I told you that I was going all the way out," he said to Rice. He was bundled up in a jacket now, steam rising from his sweaty head. "I only wish there had been good weather and a faster track — but I got no kick."

A few minutes later, after stretching and drying off, he unburdened himself to several reporters. He was asked again about Hitler, who had left just before the flag and anthem ceremony. "Immaterial," Owens said. While he said he was not bothered by the Führer's show of disrespect, he was saddened to think that his Olympics were over. There was, after all, so much he felt he still had to offer.

"I'm just getting the feel of the track," he said. "I never was in better condition and would like to keep on competing. I told Coach Robertson I'd gladly run in the relay if he wanted me. I hope there's a chance that they will keep me busy the rest of the week. I'm having a wonderful time."

19

The Relay

F ROM THE TIME of the trials through his victory at 200 meters, it had been universally assumed that Jesse Owens
would rest after the fourth day of the games and yield the Olympic spotlight to his less gifted teammates — and non-American
runners. "Jesse Owens's Olympics ended yesterday," Joe Williams wrote in his column for the August 6 *New York World-Telegram*. "Today the Olympics take on the international form for
which they were originally designed. Which is to say some of
the athletes in other countries will get a chance to break into
the headlines. Up to now it has been all Owens. The Ohio Negro has been the whole show." Responding to Owens's request
to compete in the relay, Lawson Robertson said, "Owens has had
enough glory and collected enough gold medals and oak trees
to last him a while. We want to give the other boys a chance to
enjoy the 'ceremonie protocalaire' [the victory ceremony]. Marty
Glickman, Sam Stoller, and Frank Wykoff are assured places on
the relay team. The fourth choice rests between Foy Draper and
Ralph Metcalfe."

With that, Robertson seemed to lay the issue to rest. Jesse Owens would return home with three gold medals, not four. That
was one point on which it seemed everyone agreed — everyone
except Owens himself.

In fact, while in Berlin, Owens had expressed his desire to run on the relay team several times. On August 4, for instance, Alan Gould wrote, "Jesse, who hates to stand around, had hoped to run one leg of the 400-meter sprint relay, but Lawson Robertson, track coach, feels the Ohio State Negro has done just about enough in one Olympics."

It appeared that Owens's entreaties had made no impact. Operating under the assumption that the Olympics were over for the Dark Streak from Ohio State—John Kieran's favorite nickname for Owens—Kieran wrote in the *New York Times*, "Taking the 200-meter run at Berlin yesterday, Jesse Owens made it game and set." He compared Owens, at length, to Alvin Kraenzlein, the American who had won four gold medals at the Paris games in 1900, but he never suggested that Owens might have a chance to equal Kraenzlein. "If they had wanted to make a real contest of his events they should have made Jesse run with a suitcase in each hand," Kieran wrote.

Kieran's column appeared on August 6. But on the evening of Friday, August 7, Alan Gould broke the story that Owens had been named to the relay team. Robertson "decided definitely tonight to call on Jesse Owens," he wrote. Robertson told Gould that he had changed his mind because he had received information that "the Germans quietly have built up a quartet which had been clocked in sensational time." Robertson also reportedly perceived that the Dutch might be a threat.

"That's swell news," Owens said when Robertson told him and Snyder of the shift. "I'll sure hustle around that corner." Snyder explained to the press that Robertson had made his decision in part because Owens had proved in the 200 meters how well he could run around a turn—as if that had ever been in doubt.

When Robertson told Owens and Snyder of the decision, they were appreciative, but they also felt guilty—guilty that their campaigning would cost another American a gold medal. That guilt was diminished somewhat by the reports that the Germans had a formidable foursome. Of course Owens and Snyder did not want to see the Germans win and Hitler rejoice, especially not after what *Der Angriff* had been printing. Still ... if Metcalfe did not make the team, it would mean that he would never win an Olympic gold medal; for Draper, Stoller, and Glickman, getting kicked off the relay would mean that, at best, they would have to wait another four years for gold; and for Wykoff, it would mean that he would return to the States with only a fourth place in the 100 meters. At least Wykoff had won gold medals on both the 1928 and the 1932 relay teams.

The final assignments had not been announced, but Snyder said he assumed Owens would lead off, followed by Metcalfe, Stoller, and, on the anchor leg, Wykoff. Thus it seemed that the odd men out would be Glickman and Draper, who had lagged behind Stoller in practice sprints at the athletes' village. Gould wrote that if Glickman and Draper did not receive "their expected assignments," there would be a "critical reaction among those Americans feeling that as many of the boys as possible should get to compete."

Two days earlier, Lawson Robertson had said he wanted to see as many of the so-called boys compete as possible, echoing the plan as it had been conceived back on Randall's Island. Now, like a skilled debater, he deftly took the other side. "I'd like to let everybody run, but we're here to win as many events as possible," he said. "They are likely to criticize any decision I may make, but my job's to put the best possible teams in the race." Clearly, that was not how he had previously perceived his job, or Jesse Owens — the best sprinter anyone had ever seen — would always have been on the relay team. Nor had Robertson perceived his job that way four years earlier, when Eddie Tolan and Ralph Metcalfe, his two best sprinters, both sat out the relay as four other sprinters — all white — easily won the gold.

Robertson's other stated reason for inserting Owens into his relay lineup — fear of clandestinely brilliant German and Dutch sprinters — was demonstrably absurd. Nothing the Dutch or Germans had achieved in Berlin or in the months leading up to the games suggested that they were capable of pulling off an upset. No German sprinter had won a medal in Berlin, or even placed fourth. Only Erich Borchmeyer was world-class, and that was debatable. And the Dutchman Martin Osendarp's two bronze medals hardly constituted a legitimate threat to American sprinting hegemony.

It was true too that for the 4 x 400-meter relay, which would be run the day after the 4 x 100-meter relay, Robertson did not select his four best runners; instead, he left his two best runners off that relay team. On August 7, Archie Williams won the gold medal in the 400 meters and Jimmy LuValle won the bronze medal. Arthur Godfrey Brown of Great Britain and his compatriot William Roberts finished second and fourth, respectively. Yet despite the strength of the British team — which would include both Brown and Roberts — Robertson selected for the relay team neither Williams nor LuValle, two of the eight black men on the U.S. track-and-field team. On August 9, he would pay the price for that decision. The United States lost to Great Britain by two full seconds.

So why did Robertson really change his mind? There are any number of more plausible reasons than the ones he offered. He might have felt pressure from Avery Brundage. He might have been swayed by Dean Cromwell. He might have had an honest reconsideration. But what happened on the morning of August 8, when he and Cromwell together announced to his sprinters who was in and who was out, suggests that the suspicions formed seventy years ago contain more than a morsel of truth.

That morning the seven American sprinters gathered together at the athletes' village. With Cromwell at his side, Robertson would make the announcement. By that time, everyone assumed that Owens, Metcalfe, and Wykoff were in. The question was which of the three remaining 100-meter men were out. (Mack Robinson was never considered a real possibility to run.) Glickman and Draper assumed they were but were unsure. Stoller had been told by Cromwell that he was in, but he did not entirely trust him.

"Boys, this is a tough decision," Robertson said, kneading his hands together as if he were pained to have to make a choice. "You've all done so well here, and I hate to disappoint any of you. But Coach Cromwell and I have decided to go with Jesse, Ralph, Foy, and Frank. Marty, Sam, I'm sorry, but that's our decision. We were hoping to give you both a chance, but we can't afford to take the Germans lightly. We have reason to believe that they've been hiding their best runners, waiting to upset us."

For much of the rest of his life — but not until several years later — Marty Glickman claimed that Owens stood up and said, "I've got my medals, coach. Let Marty and Sam run." At which point Cromwell supposedly replied, "You'll do as you're told," and Owens sat back down. But Owens, for his part, later reported that he stood up and said, "I've already got three gold medals, I don't need any more. I'd like to relinquish my place to Sam Stoller." In his account, Owens did not mention Glickman, which makes sense, because it seems unlikely that he would have suggested that Metcalfe should also relinquish *his* spot.

In any case, when Robertson's words hit Stoller, they hit hard. It was his twenty-first birthday. It was unlikely that he would still be sprinting when he was twenty-five.

Stoller was speechless, but Glickman stood up and spoke. Almost trembling with indignation, he said, "Coach, this is ridiculous. You don't hide world-class sprinters. They have to get race experience. Any American relay team you pick will win by fifteen yards — *any* of our runners, the milers or the hurdlers, could run against the Germans or anyone else and win by fifteen yards." He paused. "If you drop us," he continued, "there's going to be a helluva furor — we're the only two Jews on the track team."

"I'll worry about that," Robertson said.

Finally Stoller spoke. "Coach," he said, addressing Robertson, not Cromwell, "why did we go through all those races if they weren't going to mean anything? With all due respect, Foy finished last — Marty and I beat him."

"Sam," Robertson said, "Foy has more experience." Then, firmly, "That's our decision."

It is possible that Owens did stand up and protest—but barely. He had spent much of the previous week lobbying to run—not just privately, but in the press, all but demanding a chance to win a fourth gold medal, which, incidentally, was his right. He was, after all, the best runner in the world and the hero of the Olympics.

As Ralph Metcalfe recalled many years later, "Jesse is one of my best friends. I'm glad he won [his] medals, but he already had three when the relay meeting was held. But he didn't say a word that I recall. I guess he wanted number four that bad."

(On the topic of the infamous meeting, Jesse Abramson wrote in the *New York Herald Tribune*, "In justice to [Owens] it must be said that he was willing to step aside for Stoller." Again, no mention of Glickman. Of course, Abramson was not an eyewitness.)

Frank Wykoff's account of what happened is consistent with Metcalfe's—and at odds with Glickman's and Owens's. "Originally, it was definitely supposed to be Marty Glickman, Sam Stoller, Ralph Metcalfe, and myself," he later said. "Then the night before the race they announced that Jesse Owens insisted on going for four gold medals. Then I heard that Metcalfe was not going to run, that Sam Stoller would. I felt very bad. The four of us—Glickman, Stoller, Metcalfe, and I—had been working well together. We could have set the record." Like Glickman, however, Wykoff was convinced—and remained convinced—that Glickman and Stoller were the victims of anti-Semitism.

Glickman blamed not only Cromwell and Robertson but Avery Brundage as well. He theorized that the coaches and Brundage were anti-Semites and dropped him and Stoller so as not to embarrass their German hosts by sending two Jews to the top of the medal stand. Stoller, meanwhile, always said that he did not think anti-Semitism was the deciding factor. He said Cromwell was merely protecting his Trojans, Wykoff and Draper.

The protestations of Glickman and Stoller failed to move their coaches, and that afternoon at 3 P.M. Owens was on his mark in the Americans' heat. He ran a remarkable leadoff leg, handing off the baton to Metcalfe with a 20-foot lead. Thirty seconds later, Wykoff broke the tape, and the quartet of Owens, Metcalfe, Draper, and Wykoff had won its heat in 40.0 seconds, tying the world record. A few minutes later the Germans won their heat, in 41.4 seconds, and then the Dutch won their heat, in 41.3 seconds — but none of the German or Dutch sprinters were unknowns or had been hidden; they were decidedly the usual suspects and, as usual, slower than the Americans. "The heralded Dutch and German opposition, which Head Coach Lawson Robertson said prompted him to use Owens and Draper, failed to materialize," the Associated Press reported.

Climbing into the stands after the heat to watch three Americans fight it out for the decathlon championship, Owens donned a pair of sunglasses, although the sky was overcast. For the first week of the games he had enjoyed the mobs that adored him and sought him out. Now he was tired. He had run nine races and made nine jumps in eight days — and his right arm had cramped from signing his looping signature so often for so many admirers, most of them German. Larry Snyder told reporters that he feared "the nervous tension" might spread the cramp to Owens's legs. That afternoon in the stands of the Olympic stadium, Owens was hoping not to be recognized. But of course he was not hard to spot. Awestruck German fans asked him in broken English to pose with them and of course to sign autographs. He obliged, but celebrity fatigue had set in. As he tried to be good-natured, he thought about the one race that had not yet been run. He knew it would be a walkover, a certain fourth gold medal. He asked himself if he had been selfish to demand a place on the relay team. *Really,* he thought, *what's the difference between three and four gold medals?* But now it was too late. The same four men who ran in the heat were required to run in the final. He reminded himself again that he was the best in the world and it was his right to ask to run. No one could reasonably say that the United States would have a better chance of winning without him on the relay. Still, he felt guilty.

Snyder tried to assuage that guilt. "You've got no reason to feel bad about this," he said, searching for Owens's eyes behind the dark lenses. "If you were white, Robertson and Cromwell would never have considered *not* putting you on the relay. Remember, when Stoller and Glickman get home, they'll have plenty of opportunities. Glickman's only eighteen. Jesse, these are your games, you're a competitor. You deserve that gold medal tomorrow."

"I know, coach," Owens said, his head down. "I just feel bad for Sam and Marty."

Sam Stoller had been too depressed to leave the athletes' village after the morning meeting, but Marty Glickman, wearing a USA sweatshirt, spent much of the afternoon in the press box, in the company of the American writers. After watching the Germans and the Dutch prove that there had been nothing for Robertson and Cromwell to fear, he said, "The heats failed to show the necessity for shaking up the lineup after Stoller and myself long practiced the stick work. It looks like politics to us."

"How's that?" Arthur Daley asked.

"Cromwell's influence, looking out for his Trojans," Glickman said, avoiding, for the time being, any mention of anti-Semitism. "Any American combination might have run forty seconds flat this afternoon, since there was no pressure involved. I am willing to admit the team picked, at its fastest, probably can break forty for a new record, but this talk about the Germans and the Dutch being so tough looks like a false alarm."

Daley, Gallico, Gould, and the rest of the writers watched from the press box as Glenn Morris overcame his fellow Americans Bob Clark and Jack Parker to win the decathlon. At dusk, Morris concluded an exhausting two days of competition by sprinting the final few hundred meters in the final event, the 1500 meters, breaking his own decathlon world record by twenty-five points. In the fading twilight, the three Americans mounted the medalists' podium and stood at attention as "The Star-Spangled Banner" played yet again.

When Morris, a twenty-four-year-old automobile salesman from Fort Collins, Colorado, left the podium, he walked straight up to the smartly dressed older woman he had been eyeing for a week.

"I held out my hand and congratulated him," Leni Riefenstahl later wrote, "but he grabbed me in his arms, tore off my blouse, and kissed my breasts, right in the middle of the stadium, in front of a hundred thousand spectators. A lunatic, I thought."

It was undoubtedly the most dramatic moment of the games that Riefenstahl's cameramen failed to capture, and the beginning of a brief, torrid affair. (Riefenstahl said that when she received a handwritten letter from Morris and saw his "strangely convoluted strokes," she became uneasy and asked a handwriting expert for his opinion. "This is a man who is unstable," the graphologist told her. "He's dangerous, uncontrolled, ruthless, inconsiderate, brutal, and even has a sadistic streak." In 1938, Morris made the most of his animal instincts, starring as the lord of the apes in *Tarzan's Revenge*. His leading lady was also an Olympic champion: Eleanor Holm.)

For Riefenstahl, the first eight days of the games had been an unprecedented challenge. She had battled Goebbels, the IOC, and the officials on the field, all of whom had tried in vain to rein her in. She had flattered and finessed Hitler into giving her almost all that she had requested while her cameramen had shot millions of feet of footage — of runners, pole vaulters, javelin throwers, marksmen and rowers and equestrians. They had captured the full scope of the games with innovative technology and an artist's eye. There was only one problem. The star of their film-in-the-making was a black American. Riefenstahl could still foresee the problems that might pose, but she pressed on, filming Jesse Owens's every move, capturing hundreds of heroic images of him at full speed and in midair.

"What are we going to do with this man?" her chief cameraman asked her. "Goebbels will never let you release a film that celebrates *him*."

"There is no choice," she said sternly. "He is these Olympics. There can be no discussion."

"But there will be," the cameraman said, chuckling.

The next day, August 9, at 3:15 P.M., Jesse Owens was about to run his final race at the games of the Eleventh Olympiad, leading off for the United States in the final of the 4 x 100-meter relay. It was the most pleasant day he had yet experienced in Berlin. The temperature was in the mid-seventies, and the sun had emerged from captivity. The crowd was the biggest ever at an Olympics, an estimated 120,000, headlined by the Führer and his guests of honor, King Boris of Bulgaria and Crown Prince Umberto of Italy. Six teams were in the final — the Argentines on the inside, then the Germans, the Dutch, the Americans, the Italians, and the Canadians. Once more the German starter soothed Owens's nerves with his calmness and perfect timing.

At the crack of the gun, Owens tore from his mark. Despite his exhaustion, despite all the energy and emotion he had expended in the previous week, he ran faster than he had ever run before. He ran the way Charles Riley had taught him to run — his form perfect, his head up, his eyes straight ahead, his arms churning, his feet only skimming the track. This, he knew, would be his final chance to show the Germans, to show everyone, exactly how good he was. For nine seconds he tore through the red dirt. "Owens lit out as though escaping from Beelzebub," Gallico wrote. Wilhelm Leichum, the German leadoff man, and Tjeerd Boersma, the Dutch leadoff man, were 2 yards behind Owens when he handed off to Metcalfe. Also running his final Olympic race, Metcalfe only increased the American lead before handing off to Draper. Owens and Metcalfe built up such a lead, Gallico wrote, "that the white boys to whom they turned over the baton could have crawled in on their hands and knees."

Wykoff crossed the finish line in 39.8 seconds, world-record time, surprising no one. Surprising Robertson and Cromwell, perhaps, neither the Germans nor the Dutch won the silver, which went to the Italians. The Germans took bronze; the Dutch were disqualified for botching a handoff but had been no threat anyway. "The 400-meter relay was the romp everyone except Lawson Robertson expected," Jesse Abramson wrote in the *New York Herald Tribune*. "With Sam Stoller and Marty Glickman the team might not have broken the record, but these two sprinters had every right to be on the team."

When the race ended, Snyder—who had managed somehow to gain access to the field—went running to the spot on the infield where Owens had been for Wykoff. He was clutching his hat in his right hand and his program in his left. "Jesse," he said. "You did it. Man, you did it. I've never seen you run so fast."

"No reason to hold back," Owens said. Then they embraced, tears running down Snyder's cheeks.

"Come on, Larry," Owens said, "don't let the Germans see you cry."

Moments later, one last time, Owens took to the medal stand —although he insisted that Metcalfe take the highest step. He was presented with another laurel wreath and stood at attention for the flag-raising. As the Stars and Stripes was hoisted, alongside the fascist flags of the Mussolini and Hitler regimes, he caught sight of Luz Long, who had come out to the stadium to cheer him. At that moment, all his emotions washed over him in a wave—his euphoria, his pride, his guilt. Gathering himself, he looked up into the stands to the Führer's box. He thought he saw Hitler waving at him.

Epilogue

A
S JESSE OWENS was in the process of receiving his fourth
gold medal, Joe Louis and Max Schmeling were chatting
amiably with each other in Pompton Lakes, New Jersey.

Six weeks earlier they had fought twelve brutal rounds at
Yankee Stadium, Schmeling destroying the myth of Louis's invincibility. Now, as Louis trained for his fight the following week
against the former heavyweight champion Jack Sharkey, they
had their first opportunity to discuss the fight they had fought
with each other. Instead, they talked about Jesse Owens.

"Did you see the games?" Louis asked, leaning on the ropes
of the ring in which he was about to spar — poorly, it turned
out — four rounds with four different men.

"Oh, sure," Schmeling replied. He was in heavy flannels, despite the Jersey heat.

"Did you see Jesse Owens run?" Louis asked. He knew that
Schmeling had just arrived on the *Hindenburg* the night before.

Schmeling nodded slowly. "Oh, yes," he said.

Louis nodded too. "Man, can't that boy run."

"Oh, yes."

That night, with four gold medals in his suitcase, Jesse Owens
left Berlin. To cover the cost of sending its team to Germany, the
American Olympic Committee and the Amateur Athletic Union
had arranged an exhibition tour that would feature its most visi-
ble stars, who of course would receive no compensation for their
efforts. So Jesse Owens and Ralph Metcalfe, among others, were

shipped off to Cologne, a few hundred miles south and west of the capital, where 35,000 fans came out to see in person the men — most of them black — about whom they had been hearing so much. To say that the Americans were less than enthusiastic would be a huge understatement, but it would have been unacceptable to perform indifferently. Still, Metcalfe was less indifferent than Owens, defeating him for the first time in more than a year, by one tenth of a second. For his part, Owens did win the broad jump, but leaped only 24 feet, 4½ inches.

The results in Cologne were not nearly as newsworthy as Owens's announcement that he was ready to turn pro. "I'm anxious to finish college," he said to Alan Gould just before leaving Berlin, "but I can't afford to miss this chance if it really means big money. I can always go back and get a degree. It would mean giving up my future athletic career, but I have had a fair share of track-and-field honors and I feel I could hang up my spikes without any serious regrets."

In the previous week, Owens had received dozens of offers to cash in on his Olympic achievements. Eddie Cantor offered him $40,000 to appear in his act for ten weeks. An orchestra in California offered him $25,000 to spend ten weeks telling jokes as it warmed up. Wilberforce College in Ohio made him a much less lucrative offer to coach its track team, promising only "to take care of him."

Owens had one year of eligibility remaining at Ohio State, but Larry Snyder, for one, wanted to see him capitalize on his medals. "It would be foolish for me to stand in Jesse's way," he said. "He's absolutely at the height of his fame now. Nothing that he could do in his remaining year of college competition would lift him to a higher peak in the athletic world than he now enjoys. He has a good chance to make $75,000 to $100,000. I'd be glad to see him do it."

Snyder was right. In August 1936, Owens was as famous as any man in America — praised equally by the white and black press, celebrated for refuting Hitler's claims of Aryan superiority. But then there was the reality of his status as a black man in the United States — and as an amateur athlete competing at the pleasure of a man who, as his power increased, would come to be known as Slavery Avery.

After Cologne, Owens participated in exhibitions in Prague and in Bochum, Germany, flying around Europe with so little cash that he was often fed by the strangers seated next to him. Still the offers kept coming. By August 15, he and Snyder were in London for another AAU event. There he showed a reporter the telegram from Cantor, Will Rogers's one-time colleague in the Ziegfeld Follies. "I don't know what I'm going to do," Owens said, reading the telegram for the hundredth time. "All I know is that I want to get back to my family. I'm going to Sweden Monday, then back to London, then home — and I'll be the gladdest man in the world to get there, believe me. All this ballyhoo is getting on my nerves."

Owens went out and ran the third leg in the sprint relay, this time run at the nonmetric distance of 440 yards, and this time with Ralph Metcalfe, Frank Wykoff, and Marty Glickman. They set a world record, but Owens was done. After the race he told Larry Snyder that that was it. There would be no trip to Stockholm. They were going home as soon as possible.

The next day, back in Berlin, on the final day of the games in which Jesse Owens had won four gold medals, set record after record, and upheld the honor of the United States in the face of Nazism, Avery Brundage indefinitely suspended him for refusing to proceed to Stockholm, making him ineligible to compete in AAU-sponsored events. "We had no alternative under the circumstances but to disbar Owens," Daniel Ferris, the AAU secretary, said at the Olympic stadium.

Snyder, still in London, was apoplectic. "Last Sunday," he said, his face reddening, "Jesse was informed by Ferris just forty-five minutes before time to leave that he was going to compete in a meet at Cologne. This was Sunday night, after Jesse had won the relay for the United States following a tough week of competition. He had to hurry and pack one suitcase, leaving the rest of his things for Dave Albritton to pack. The day after he met Ralph Metcalfe [in the 100 meters]. It looks to me like a deliberate attempt to have the boy beaten. It was absolutely unfair. It doesn't matter whom the AAU sacrifices to get its 10 percent. You wouldn't ask the poorest show troupe to work the way these boys worked immediately after the games — all without a cent of spending money with which to brighten an otherwise drab picture."

"There's nothing I can gain out of this trip," Owens said when Snyder finally stopped talking to take a breath. "This suspension is very unfair to me. All we athletes get out of this Olympic business is a view out of a train or airplane window. It gets very tiresome, it really does."

Then, his own ire rising nearly to the level of Snyder's, he added, "This track business is becoming one of the great rackets in the world. It doesn't mean a thing to us athletes. The AAU gets the money. It gets all the money collected in the United States and then comes over to Europe and takes half the proceeds. A fellow desires something for himself."

In Cleveland, reporters knocked on Ruth Owens's door. She was at home with Gloria and her mother-in-law.

"Mrs. Owens, what do you think of this suspension?" the man from the *Plain Dealer* asked.

"It is pretty terrible," Ruth said. Then Emma Owens waved her hands. "Terrible isn't a word for it," she said, cutting off Ruth. "It's *scandalous*. Hasn't my boy done everything that was asked of him? Didn't he run his legs off to bring victory to the team? If he fails to get the money now, he'll simply be a forgotten Owens in ten years."

For three days Snyder and Owens sat around in their hotel in London, mulling over the offers to appear on Broadway, or with Bojangles, or with Eddie Cantor. Finally, on Wednesday, August 19, they boarded the *Queen Mary* for a weeklong voyage home. Ruth went to New York with Jesse's parents to see her husband for the first time in two months — but together they spent a frustrating and humiliating night being rejected for service by hotel after hotel. Finally the Hotel Pennsylvania gave them rooms — on the condition that they use the service entrance. Even in New York, it didn't matter whether you were the world's greatest athlete, if you were black.

Not surprisingly, the promises of easy wealth were all lies. No one was actually willing to pay Owens tens of thousands of dollars to do *anything*. All those telegrams — with the exception of the offer from Wilberforce — were publicity stunts, designed to place names in newspaper columns.

Eventually Owens tried to cash in on his fame with a chain of dry-cleaning stores. They failed. He owned and operated a barnstorming black baseball team. To attract crowds to the games, he sometimes raced a horse across the outfield. Of course the horse would spot him 20 yards — and Owens, sometimes in a suit, sometimes in shorts, would sometimes win. If the horse got a bad start.

He took a job with the state government of Illinois, working as a kind of physical education guru in the school system. Here he excelled, sharing the lessons he had learned from Charles Riley and Larry Snyder with thousands of children. He traveled the world spreading the Olympic gospel, mostly leaving to Ruth the responsibility of raising their three daughters. He also worked as an executive at Ford Motor Company and with a sporting goods company. All the while he maintained close friendships with Larry Snyder, who spent his entire career at Ohio State and served as the head coach of the 1960 U.S. Olympic track team; Ralph Metcalfe, who eventually served four terms in the U.S. House of Representatives; Marty Glickman, a pioneering sports broadcaster; and even Eulace Peacock, a sometime business partner.

Owens's friendship with Luz Long was not as enduring. For several years the two athletes maintained a correspondence, sharing their troubles and their hopes, but shortly after receiving his law degree, Long was compelled to join the German armed forces. By the time he wrote the following letter to Owens, the United States and Germany were at war, and he was too:

> My heart is telling me that this is perhaps the last letter of my life. If that is so, I beg one thing from you: When the war is over, please go to Germany, find my son and tell him about his father. Tell him about the times when war did not separate us — and tell him that things can be different between men in this world.
>
> Your brother,
> Luz

One of Hitler's more reluctant soldiers, Luz Long was fatally wounded during the Allied invasion of Sicily. He died in a British field hospital on July 13, 1943, at the age of thirty. In 1951, Jesse Owens did what Long had asked him to do. He found Kai Long and told him about his father. "I've seen Luz again," he said, "in the face of his son."

In the 1960s, as the civil rights movement turned increasingly militant, black radicals took to calling Owens an Uncle Tom. After John Carlos and Tommie Smith raised their black-gloved fists on the 200-meter medal stand at the 1968 Olympics in Mexico City, Owens — hoping to control the damage — met with the U.S. track team. The black athletes all but spat in his face.

Then, in the early 1970s, he embraced black militancy, publishing *I Have Changed,* a book that reflected the evolution of his political philosophy. Still he continued to speak at corporate gatherings, to the Elks and the Rotarians and the Boy Scouts. He became, in the words of William O. Johnson of *Sports Illustrated,* "a professional good example," sharing his stories of Adolf Hitler and Luz Long, of Charles Riley and Larry Snyder. Owens gave his audiences what he thought they wanted — more than anything else, the false impression that he had been snubbed by the Führer and that the German crowds had been hostile to him and his black teammates during his fortnight in Berlin. In his mind, he easily justified his dissembling. Denied by white America the opportunities for wealth that he thought he was owed, he exaggerated his stories to make a good living. So what?

In 1980, at the age of sixty-six, James Cleveland Owens died of lung cancer. In his obituaries, he was called, unfailingly, the greatest of all Olympic stars. Even today, more than 110 years after the revival of the games, he remains their ultimate champion.

Leni Riefenstahl spent two years editing *Olympia*. When it was finally released, in 1938, critics hailed it as one of the greatest achievements of the cinema. *Olympia* remains a staple of film schools, and its technical and stylistic innovations have influenced generations of filmmakers. Riefenstahl did have to fight Goebbels to keep all her beautiful shots of Owens, who emerges from the film as the god of the games. His beauty and grace were a rebuke to the regime that Riefenstahl had done so much to glorify.

The most spectacular image of Owens in *Olympia* shows him making his last attempt in the broad-jump competition, the jump that stood as an Olympic record for twenty-four years. Actually, like much of *Olympia*, the shot is a re-creation, with Owens jumping just for Riefenstahl's camera. First he is seen from the side, in his famous, revised starting position. Then he is hurtling down the track, and as he hits the takeoff board, Riefenstahl cuts to a head-on shot. The camera stays with him as he leaps several feet off the ground, folds his legs, stretches his arms straight ahead, and then crashes into the dirt. Then Riefenstahl cuts to the actual jump. Owens leaps up from the pit, sand clinging to his dark legs. Then he is standing facing the camera, a disembodied white arm on his shoulder. On his face there is a smile of deep satisfaction. He is serene and handsome.

Riefenstahl's message is clear: Look closely. Here is your superman.

Notes

8 In the previous day's: Francis J. Powers, "Jesse Owens Can Leap 27 Feet, Says Coach," *Los Angeles Times,* May 24, 1935, p. A17.

9 "by about a foot": Heisler, "A Day They Will Not Forget."

10 "absolutely beautiful": Ibid.
     "Every coach in the Big Ten": Snyder, "My Boy Jesse."

12 "It was like an accordion": Heisler, "A Day They Will Not Forget."

13 "I broke out": Jesse Owens, *I Have Changed* (New York: William Morrow, 1972), p. 32.
     "I want to tell you": Alvin Silverman, "Owens' Words Tell Why He Rose from Junior High 'Punk' to 'Fastest Human,'" *Cleveland Plain Dealer,* Aug. 6, 1936.

2. OUT OF ALABAMA

14 "My father's long": Jesse Owens, *Jesse: The Man Who Outran Hitler* (Plainfield, N.J.: Logos, 1985), p. 37.

17 "We've got to do something": Ibid., p. 5.

18 Between 1882 and 1902: The History of Jim Crow, http://www.jim-crowhistory.org/scripts/jimcrow/glossary.cgi.
     "I'm going to take": Owens, *Jesse,* p. 6.

19 "We never had any problems": Ibid., p. 10.

20 "I always loved running": Ibid., p. 18.
     In 1922, when Jesse: William J. Baker, *Jesse Owens: An American Life* (New York: Free Press, 1988), p. 16.
     "It's crazy to go on": Owens, *Jesse,* p. 22.

21 In Cleveland, J.C. enrolled: Baker, *Jesse Owens,* p. 19.
     One day when Jesse: Bob Dolgan, "Cleveland's Golden Son," *Cleveland Plain Dealer,* Sept. 25, 2000, p. 2C.

23 "I thought I'd win": Owens, *Jesse,* p. 39.

24 "Don't talk, Jesse": Ibid., pp. 40–41.
     Soon thereafter Jesse found: Ibid., pp. 43–44.

25 Four years later, in Paris: The 200-meter bronze medalist was Eric Liddell, a Scottish missionary whose story was chronicled in the Academy Award–winning film *Chariots of Fire.* Paddock failed to place in 1924 at 100 meters, the race that was won by the Englishman Harold Abrahams, whose story was also central to *Chariots of Fire.*

27 "I haven't got the heart": Alvin Silverman, "Owens' Words Tell Why He Rose from Junior High 'Punk' to 'Fastest Human,'" *Cleveland Plain Dealer,* Aug. 6, 1936.

29 "hunted up interpreters": "Sexton Cracks Shot-Put Mark," *Los Angeles Times*, Aug. 22, 1932, p. 7.
"He will be an asset": Baker, *Jesse Owens*, p. 35.
In just one year: "Soccer Outfit Will Not Quit," *Los Angeles Times*, Dec. 25, 1932, p. D1.
He also made: "Ohio State to Train Runners in New Style," *New York Times*, Sept. 3, 1932, p. 10.

3. VINCIBLE

31 "Dear old Babe Ruth": Will Rogers, "Mr. Rogers Spent Sunday with the Sporting Pages," *New York Times*, May 26, 1935, p. 19.

32 "a Republican holiday": Will Rogers, "Mr. Rogers Takes Note of the NRA Slaughter," *New York Times*, May 28, 1935, p. 27.

33 "I think the praise": William J. Baker, *Jesse Owens: An American Life* (New York: Free Press, 1988), p. 52.

34 "After last Saturday's performances": "Jesse Owens Recalls Snub by Metcalfe," *Chicago Daily Tribune*, May 29, 1935, p. 22.

35 "most of [them]": Wilfrid Smith, "Owens Leads Ohio State to Central Track Title," *Chicago Daily Tribune*, June 8, 1935, p. 21.
"almost heartbroken": "Jesse Owens Recalls Snub by Metcalfe."

36 "There was no doubt": Ibid.

37 "under-cover of the semi-darkness": Bill Henry, "Ohio State Due Today," *Los Angeles Times*, June 11, 1935, p. A13.
"Perhaps the Indians": Harry Carr, "The Lancer: The Negro Wonder," *Los Angeles Times*, June 11, 1935, p. A1.
"a good-natured colored boy": Bill Henry, "Jesse Owens Faces Lens," *Los Angeles Times*, June 12, 1935, p. 7.
"The local colored colony": Bill Henry, "Bill Henry Says — ," *Los Angeles Times*, June 14, 1935, p. A15.

38 "Setting here": Will Rogers, "Mr. Rogers Is Hobnobbing with Some Fast People," *New York Times*, June 13, 1935, p. 25.
Jim Thorpe was working: Sadly, *Steamboat Bill* was Rogers's final picture. Two months after Owens visited him, Rogers and his friend Wiley Post were killed near Point Barrow, Alaska, when their plane's engine failed on takeoff. Also an Oklahoman, Post was one of the celebrity aviators of the era and had twice flown around the world. "President and pauper alike expressed sorrow, for both men were known over virtually the length and breadth of civilization," the Associated Press reported from Alaska.

39   "It was good for him": Larry Snyder, "My Boy Jesse," *Saturday Evening Post*, Nov. 7, 1936, p. 15.

"Short muscles in the calf": Braven Dyer, "Why Is Jesse Owens Greatest Track Star?" *Los Angeles Times*, June 13, 1935, p. A11.

40   "Nobody can answer that": Ibid.

"Don't miss today's appearance": Braven Dyer, "The Sports Parade," *Los Angeles Times*, June 15, 1935, p. 9.

"He won so easily": Bill Henry, "Owens Wins Four Events but Troy Triumphs," *Los Angeles Times*, June 16, 1935, p. 23.

41   "I never saw": Associated Press, "Frank Wykoff Marvels at Ease of Jesse Owens' Track Form," *Washington Post*, June 17, 1935, p. 15.

"I have to get my lead": Ibid.

43   "I feel fine": Bill Henry, "'I Was Lucky to Win Four Firsts,' Owens," *Los Angeles Times*, June 24, 1935, p. 10.

"trailed Owens": Associated Press, "Jesse Owens Dances after Taking 4 Events," *Washington Post*, June 24, 1935, p. 16.

"He beat just the pick": "Owens Enters Four Events, Wins Four National Titles," *Chicago Daily Tribune*, June 23, 1935, p. A1.

44   "There were only nine inches": Braven Dyer, "Owens Double Winner at San Diego," *Los Angeles Times*, June 28, 1935, p. A11.

45   "Jesse Owens, Negro college student": "A Solomonic Judgment: The Submerged Tenth," *Los Angeles Times*, June 28, 1935, p. A4.

"So important": Baker, *Jesse Owens*, p. 57.

46   "one last chance": *Cleveland Plain Dealer*, "Owens, Told He'd Be Sued, Will Sprint to Wed Maid He Wooed," July 4, 1935.

"Possibly Ralph Metcalfe": John Kieran, "Sports of the Times: Just Jesse, a Tale of Speed," *New York Times*, June 30, 1935, p. S2.

"Fortnight ago": "Negroes in Nebraska," *Time*, July 15, 1935.

47   "However," he said: Donald McRae, *Heroes Without a Country: America's Betrayal of Joe Louis and Jesse Owens* (New York: Ecco, 2002), p. 82.

Finally, on their twelfth try: Arch Ward, "Talking It Over," *Chicago Daily Tribune*, July 5, 1935, p. 17.

48   "This was to have been": Arthur J. Daley, "Owens Beaten Twice by Peacock as New York A.C. Retains U.S. Track Title," *New York Times*, July 5, 1935, p. 16.

"Enter Mistah Owens": Braven Dyer, "The Sports Parade," *Los Angeles Times*, July 6, 1935, p. 7.

50   A romantic Jesse was not: Baker, *Jesse Owens*, p. 58.

"The wedding set": "Dan Cupid Catches Up with Owens," *Los Angeles Times*, July 6, 1935, p. 5.

51 "Then," Arthur Daley wrote: Arthur J. Daley, "Two Records Are Shattered in All-Star Track and Field Meet at Ohio Field," *New York Times,* July 10, 1935, p. 25.

52 "It looked to me": Paul Gallico, "Give Owens a Rest, He'll Beat Peacock in the Dash," *Washington Post,* July 11, 1935, p. 19.
"I just mean": Ibid.
"I can only see Peacock": Paul Zimmerman, "Paddock Sees Peacock as Olympic Champion," *Los Angeles Times,* July 10, 1935, p. 5.

53 "Peacock is the fastest": McRae, *Heroes Without a Country,* pp. 86–87.
"Of course I'm worn down": Ibid., p. 87.

54 "It's going to take": Ibid.
"If both men were": Gallico, "Give Owens a Rest."
"a high tension": Dick Schaap, *An Illustrated History of the Olympics* (New York: Knopf, 1975), p. 210.
"Owens is a form runner": Ibid.

4. HEEL BONES AND A NEW START

55 "The latter clause": William J. Baker, *Jesse Owens: An American Life* (New York: Free Press, 1988), p. 60.

56 "industry, training, incentive": "Jesse Owens Success Analyzed by Doctor," *Los Angeles Times,* July 28, 1935, p. 18.

57 "a tempest in a teapot": Paul Zimmerman, "Paddock Takes Up Cudgel on Behalf of Jesse Owens Against A.A.U. Officials," *Los Angeles Times,* Aug. 15, 1935, p. 12.
"We failed to find": "Owens Case Is Closed," *New York Times,* Sept. 1, 1935, p. S8.
"Humanity demands": "Negro Olympic Ban Urged," *New York Times,* Aug. 23, 1935, p. 9.

58 "Owens Not Washed Up": Edward J. Neil, "Owens Works on New Start," *Los Angeles Times,* Aug. 4, 1935, p. 24.

59 "Tests so far": "New Start for Owens," *Los Angeles Times,* July 31, 1935. p. 9.

5. THE JUDGE AND THE MILLIONAIRE

63 "No better place": William O. Johnson, *All That Glitters Is Not Gold: The Olympic Games* (New York: Putnam, 1972), p. 78.

64 "You didn't have to": Ibid.
"He was sincere": Red Smith, "The Noblest Badger of Them All," *New York Times,* May 12, 1975, p. 48.

64 "There is no room": "May Ask New Pledges: Olympic Committee Members Are Alarmed by Berlin Outbursts," *New York Times*, July 26, 1935, p. 9.

65 "The German Olympic Committee": John MacCormac, "Reich Keeps Games, Giving Way on Jews," *New York Times*, June 8, 1933, p. 1.

66 "We shall see": "German Jews Face Exclusion from Olympics Despite Pact," *Washington Post*, Aug. 6, 1933, p. 21.

"all special facilities," "German Jews Barred from Olympic Games," *Los Angeles Times*, Aug. 6, 1933, p. D1.

67 "The lamentable events": Associated Press, "Toscanini Refuses to Go to Baireuth," *New York Times*, June 6, 1933, p. 1.

"The Olympic games will not": Ibid.

68 "set the seal": United States Holocaust Memorial Museum, www.ushmm.org/museum/exhibit/online/olympics/zcc034.htm.

"I think that one": Barbara S. Kraft, "Moscow's Olympic Flame Flickers in the Cold Memory of Berlin, 1936," *Los Angeles Times*, Feb. 24, 1980, p. E1.

"Now that it is": Stan Cohen, *The Games of '36* (Missoula, Mt.: Pictorial Histories, 1996), p. 3.

69 In 1930, on the occasion: "Gets German Decoration," *New York Times*, Aug. 19, 1930, p. 7.

"The issue becomes": "Nazis Seek to Oust 1936 Olympic Head," *New York Times*, Apr. 4, 1933, p. 13.

70 "To my mind": Arthur J. Daley, "Berlin Faces Loss of Olympic Games," *New York Times*, Apr. 18, 1933, p. 18.

"The German committee": Associated Press, "Brundage's Approval of Berlin's Conduct Forecasts U.S. Acceptance of Olympic Bid," *New York Times*, Aug. 11, 1934, p. 10.

71 "are an international event": Kraft, "Moscow's Olympic Flame Flickers."

"Certain Jews must understand": Smith, "The Noblest Badger."

"Should the Games": Duff Hart-Davis, *Hitler's Games: The 1936 Olympics* (New York: Harper & Row, 1986), p. 76. Messersmith, incidentally, had achieved brief notoriety in 1932 when Albert Einstein criticized American consular officials in Berlin—not Messersmith, but his underlings—who had interviewed him when he was applying for a visa to the United States. "I suggest in the future Consuls put pins in their victims' chairs so they will feel stuck from the beginning," Professor Einstein said.

72 "The fact that no Jews": "Brundage Favors Berlin Olympics," *New York Times,* July 27, 1935, p. 2.

73 "As I wished": Richard Mandell, *The Nazi Olympics* (Urbana: University of Illinois Press, 1987). pp. 50–51.

"welcomed the allotting": Organisationskomitee Für Die XI. Olympiade Berlin 1936 E. V., *The XIth Olympic Games, Berlin, 1936, Official Report,* Vol. 1 (Berlin: Wilhelm Limpert, 1937), p. 47.

74 "This sacred site": Frederic Spotts, *Hitler and the Power of Aesthetics* (New York: Overlook, 2004), p. 71.

"The stadium must": Hart-Davis, *Hitler's Games,* p. 47.

75 "could not take place": Albert Speer, *Inside the Third Reich* (New York: Touchstone, 1997), p. 80.

"Overnight": Ibid.

"In 1940, the Olympic Games": Ibid, p. 70.

77 "The Chancellor is taking": Hart-Davis, *Hitler's Games,* p. 13.

78 It was Hitler: Spotts, *Hitler and the Power of Aesthetics,* p. 354.

"I will advance": "Olympics in Germany in '36 Ok'd by Hitler," *Washington Post,* Mar. 17, 1933, p. 13.

"a veritably historic opportunity": Genet, "Berlin Letter," *The New Yorker,* Aug. 1, 1936.

79 "There has never been": Frederick T. Birchall, "All Considerations of Cost Set Aside in German Preparations for Olympics," *New York Times,* July 20, 1936, p. 19.

80 "The Negro in the South": Michael Mirer, "Ben Johnson '39: How Fast Can a Man Run 60 Yards?" *Columbia Spectator,* Jan. 19, 2000, www.columbiaspectator.com.

81 Johnson — who: Arthur R. Ashe, Jr., *A Hard Road to Glory* (New York: Amistad, 1993), p. 33.

in August 1935: William J. Baker, *Jesse Owens: An American Life* (New York: Free Press, 1988), p. 65.

82 "No American Participation": Cohen, *The Games of '36,* p. 4.

"With the possible exception": George S. Schuyler, "Letter to the Sports Editor," *New York Times,* July 11, 1936, p. 8.

6. "WE ARE WITH YOU, ADOLF"

83 "I went to Germany": "Sherrill Rebuffs Olympic Ban Plea," *New York Times,* Oct. 22, 1935, p. 1.

84 "The issue is not Germany": "Sherrill Assailed for Olympic Stand," *New York Times,* Oct. 23, 1935, p. 9.

84 That night, in Yonkers: "Sherrill Rebuffs Olympic Ban Plea."

85 "Germans are not discriminating": "Sherrill Assailed for Olympic Stand."

Rubien must have known: International Jewish Sports Hall of Fame, www.jewishsports.net/medalists.htm.

"absolutely no discrimination": "Olympic Aide Saw No Discrimination," *New York Times*, Sept. 10, 1935, p. 10.

86 DRIVE CAREFULLY!: William L. Shirer, *The Nightmare Years, 1930–1940* (Edinburgh: Birlinn, 2001), p. 230.

"his good offices": "Jahncke Asks Ban on Olympic Games," *New York Times*, Nov. 27, 1935, p. 2.

87 "The boycott campaign": Ibid.

"It is logical to expel": "Nazi Views on Sports Cited by Churchman," *New York Times*, Dec. 16, 1935, p. 10.

88 "Have no doubt of it": "Gov. Earle Urges U.S. Olympic Ban," *New York Times*, Dec. 4, 1935, p. 26.

"the regimentation of all": Ibid.

"ask for no quarter": Ibid.

"I came here": Ibid.

89 "Will you convey": "Negroes Decry Olympics, Association Urges AAU to Boycott Berlin Games," *New York Times*, Dec. 7, 1935, p. 7.

90 "No political situation": "Negro Athletes for Olympics," *New York Times*, Dec. 4, 1935, p. 26.

"These colored athletes": "17 Coaches Protest Nazi Sport Tactics," *Chicago Daily Tribune*, Dec. 3, 1935, p. 23.

91 "In retrospect": David Margolick, *Beyond Glory: Joe Louis vs. Max Schmeling, and a World on the Brink* (New York: Knopf, 2005), p. 118.

92 "What would old Adolf": Paul Gallico, "Gallico Thinks Schmeling Is Disappointed in Louis in Workout," *Washington Post*, Dec. 9, 1935, p. 20.

93 "The principal effect": *New York Times*, Dec. 7, 1935, p. 1.

"If necessary": Ibid.

"It was double-crossing": "A.A.U. Blocks Vote on Olympics Ban," *New York Times*, Dec. 8, 1935, p. 1.

94 "I am glad": Ibid.

95 "One of the immediate tasks": "Gen. Sherrill Suggests Reds Back Games Foes," *Washington Post*, Dec. 9, 1935, p. 19.

"What you are trying here": "AAU Backs Team in Berlin Olympics," *New York Times*, Dec. 9, 1935, p. 1.

96 "A famous spell-binder": Ibid.

"I bow to the will": Ibid.

7. A BLESSING IN DISGUISE

98 "If there is discrimination": William J. Baker, *Jesse Owens: An American Life* (New York: Free Press, 1988), p. 65.
"Jesse Owens is sitting": Ibid., pp. 65–66.

99 "I see no reason": Associated Press, "Owens Is Center of New Wrangle Over Olympics," *Chicago Daily Tribune*, Nov. 11, 1935, p. 22.

100 "the world's outstanding": Susan D. Bachrach, *The Nazi Olympics: Berlin 1936* (Boston: Little, Brown, 2000), p. 58.
"My dear Mr. Owens": David K. Wiggins and Patrick B. Miller, *The Unlevel Playing Field: A Documentary History of the African American Experience in Sport* (Urbana: University of Illinois Press, 2005), pp. 164–65.

102 "contributed the most": "Owens Off Honor Roll," *Los Angeles Times*, Dec. 8, 1935, p. 17.

103 "I am disappointed": "Owens, Track Champion, Ruled Ineligible," *Chicago Daily Tribune*, Dec. 29, 1935, p. A1.

104 "We regret having Owens out": Associated Press, "Owens' Enforced Rest Seen Boon to Negro's Chances in Olympics," *Washington Post*, Feb. 13, 1936, p. X16.

8. JEW KILLS NAZI

105 "Too many S.S. troops": William L. Shirer, *Berlin Diary: The Journal of a Foreign Correspondent, 1934–1941* (Baltimore: Johns Hopkins University Press, 2002), p. 46.

106 "So far as his appearance": Westbrook Pegler, "Fair Enough: Hitler at Garmisch," *Washington Post*, Feb. 21, 1936, p. 9.

107 "Nazi Germany, from the bier": Associated Press, "Nazi Leaders Move to Block Threat of Riots; Jewish Groups Forbidden to Meet After Gustloff Assassination," *Washington Post*, Feb. 6, 1936, p. 9.
"the Gustloff murder": Associated Press, "Student Slays Nazi Chieftain in Switzerland," *Washington Post*, Feb. 5, 1936, p. 1.
"a nervous, hollow-eyed young Jew": "Jew Kills Nazi," *Time*, Feb. 17, 1936.

108 "Still thinking": Ibid. Later in 1936, Hitler named a cruise ship after Gustloff. On January 30, 1945, it was plying the nearly frozen Baltic Sea with 10,000 passengers, the vast majority of them Germans fleeing the advancing Red Army, when it was sunk by a Soviet submarine. Fewer than 1250 people are thought to have survived. The sinking and the drowned are memorialized in Günter Grass's *Crab-*

walk, the Nobel laureate's 2003 novel focusing on German victimization, which also tells the stories of Gustloff and Frankfurter. Frankfurter was released from jail in 1945, just after the war in Europe ended, and moved to Israel, where he eventually served in the defense ministry. He died there in 1982.

"our Jewish enemy": Sigrid Schultz, "Hitler Pledges Unceasing War Against Jews," *Chicago Daily Tribune,* Feb. 13, 1936, p. 1.

"It seemed a fine acknowledgement": Westbrook Pegler, "Fair Enough: Olympics' Usefulness," *Washington Post,* Feb. 27, 1936, p. 9.

109   "There was a temptation": Westbrook Pegler, "Fair Enough: Entering Germany," *Washington Post,* Feb. 15, 1936, p. 7.

"Your correspondent caught": Paul Gallico, "Send Over Joe Louis," *Washington Post,* Feb. 13, 1936, p. X16.

110   Together, he and Albritton: William J. Baker, *Jesse Owens: An American Life* (New York: Free Press, 1988), p. 67.

111   "Jesse Owens has served notice": Associated Press, "Layoff Fails to End Reign of Owens," *Washington Post,* Mar. 23, 1936, p. 18.

9. A FRIEND AND A FOE FELLED

116   "A look of pain": Arthur J. Daley, "Manhattan, Texas and Ohio State Relay Teams Win Titles at Penn Carnival," *New York Times,* Apr. 25, 1936, p. 10.

118   "When Owens was smashing": Francis J. Powers, "Owens Seeks Four Wins," *Los Angeles Times,* May 20, 1936, p. A11.

119   "should give Owens": Ibid.

10. OLYMPIC TRIALS

122   "If you go": Robert A. Caro, *The Power Broker: Robert Moses and the Fall of New York* (New York: Vintage, 1975), p. 441.

123   "Fully half of the men": "Cheering Crowds Greet Roosevelt," *New York Times,* July 12, 1936, p. 22.

"Many of you": "Texts of Addresses by Roosevelt, Lehman and Others at Bridge Ceremony," *New York Times,* July 12, 1936, p. 23.

124   "This may be the best": John Kieran, "Sports of the Times: From the Bridge to the Boat," *New York Times,* July 11, 1936, p. 8.

"Athletes and sport": Westbrook Pegler, "Fair Enough: Ballyhoo at Olympics," *Washington Post,* July 11, 1936, p. X7.

125   "Germany was awarded": Ibid.

127 "Everybody up": Marty Glickman and Stan Isaacs, *The Fastest Kid on the Block: The Marty Glickman Story* (Syracuse, N.Y.: Syracuse University Press, 1996), p. 9.

128 "as if they were": Ibid., p. 10.
"Now here's Marty Glickman": Ibid.

129 "machine-like": Alan Gould, "Owens, Metcalfe and Wykoff— Down the Blazing Stretch of 100 Meters to a Boat Ride and the Olympic Games!" *Los Angeles Times*, July 12, 1936, p. A9.
"a scarlet comet": Arthur J. Daley, "Metcalfe 2d in Sprint," *New York Times*, July 12, 1936, p. S1.
"the great colored athlete": "Owens Captures Two Events in Olympic Finals," *Chicago Daily Tribune*, July 12, 1936, p. A1.

130 Everyone assumed: In the five Olympic 4 x 100-meter relays through 1932, only twice, in 1920 and 1928, did any of the top three American 100-meter runners take part in the relay. The American procedure was clearly established. Owens and Metcalfe would not take the baton in Berlin, and Frank Wykoff, Foy Draper, Marty Glickman, and Sam Stoller would.

131 "The great track and field meet": John Kieran, "Sports of the Times: Bound for Berlin, via Randalls Island," *New York Times*, July 13, 1936, p. 21.
"The Negro race's triumphs": Shirley Povich, "This Morning . . . ," *Washington Post*, July 13, 1936, p. 15.
"Jesse, my name's Jimmy Cannon": Jimmy Cannon, "Cannon Finds Owens Just a Country Boy," *New York Journal*, July 15, 1936, p. 19.

132 Mark O'Hara of the *Daily Worker:* Mark O'Hara, "The Fastest Human," *Sunday Worker,* Aug. 2, 1936, p. 2.

133 "The Nazis face": Westbrook Pegler, "Fair Enough," *New York World-Telegram*, July 16, 1936, p. 17.

II. OLYMPIA

137 In typically dramatic fashion: Leni Riefenstahl, *Leni Riefenstahl: A Memoir* (New York: St. Martin's, 1992), pp. 168–69.

138 "Riefenstahl is an actress": Mordaunt Hall, "A Drama of the Alps," *New York Times*, Nov. 29, 1927, p. 31.
Despite the success: Riefenstahl, *Leni Riefenstahl*, p. 146.

139 "The point is that Hitler": Ibid., p. 149.
Her first conversation: Ibid., pp. 178–80.

141 "These arrangements": "Nazis Will Guide Olympic Guests," *New York Times*, Apr. 26, 1936, p. 27.

12. THE BELLE OF THE BALL

142 "You gonna win": William J. Baker, *Jesse Owens: An American Life* (New York: Free Press, 1988), pp. 71–72.

143 He felt especially nauseated: Jesse Owens, diary, July 18 and 19, 1936, Ohio State University Archives.

144 "She won her place": Paul Gallico, *A Farewell to Sport* (New York: Knopf, 1938), p. 254.

145 "The campaign of vindictiveness": Ibid., p. 255.

"The Greeks had the right idea": Grantland Rice, "Eleanor Breaks Rules, Records Equally Well," *Los Angeles Times*, July 25, 1936, p. 13.

146 "has been one of the few": Ibid.

"He is very sensitive": Westbrook Pegler, "Fair Enough," *New York World-Telegram*, Aug. 1, 1936, p. 18.

147 "The smug gentlemen": Fred Farrell, "Fanning with Farrell," *Daily Worker*, Aug. 2, 1936, p. 14.

"Hitler asked me himself": William O. Johnson, *All That Glitters Is Not Gold: The Olympic Games* (New York: Putnam, 1972), p. 185.

"I had a mold": Ibid., p. 188.

148 "I am willing": "Snyder Wins Argument; to Coach Jesse Owens," *Los Angeles Times*, July 24, 1936, p. A9.

149 "You can tell": Ibid.

13. THE BATTLE TENT OF SOME GREAT EMPEROR

150 "the idol of the nation": William L. Shirer, *The Nightmare Years, 1930–1940* (Edinburgh: Birlinn, 2001), p. 236.

151 "Lindbergh proceeded to tell us": Ibid., p. 237.

152 "George observed that": Thomas Wolfe, *You Can't Go Home Again* (New York: Harper Perennial, 1998), p. 589.

153 "Hitler has got away": William L. Shirer, *Berlin Diary* (New York: Knopf, 1941), p. 55.

"From one end of the city": Wolfe, *You Can't Go Home Again*, p. 590.

Wolfe could not have known: International sports events have long been excuses for jailing undesirables and exercising authority. In 1968, before the games of the Nineteenth Olympiad in Mexico City, the Mexican government brutally suppressed a student protest movement; in 1974, on the eve of the Muhammad Ali–George Foreman world heavyweight championship fight in Kinshasa, Zaire, Mobutu Sese Seko arrested hundreds of suspected thieves and political enemies; and even now there are reports that in Beijing, the site of the 2008 summer games, the government is destroying neighborhoods that have been bastions of dissent while claiming that it is merely making necessary improvements for the Olympics.

"All anti-Semitic posters": "Nazis Put Up Fake Front for Olympics," *Daily Worker*, Aug. 12, 1936.

"Just as we breed": Jeremy Laurance, "A Short Step from Different to Undesirable," *Independent* (London), Aug. 30, 1997, p. 11.

154 "No action against": Shirer, *The Nightmare Years*, p. 232.

"no comments should be made": United States Holocaust Memorial Museum, "The Facade of Hospitality," www.ushmm.org/museum/exhibit/online/olympics/zcd059.htm.

155 "I have an idea": Fred Farrell, "Fanning with Farrell," *Daily Worker*, Aug. 2, 1936, p. 14.

"This is the most loathsome": Victor Klemperer, *I Will Bear Witness* (New York: Modern Library, 1999), p. 183.

156 "Your wandering correspondent": Paul Gallico, "Germans Play at War Near Olympic Village," *Los Angeles Times*, July 18, 1936, p. 16.

157 "Jesse Owens and his brown-skinned": R. Walter Merguson, "Race Athletes Not Involved in Olympic Scandal Charges," *Pittsburgh Courier*, Aug. 1, 1936, p. 1.

158 "steaks, and plenty of it": Donald McRae, *Heroes Without a Country: America's Betrayal of Joe Louis and Jesse Owens* (New York: Ecco, 2002), p. 143.

14. THE YOUTH OF THE WORLD

159 When it was made clear: Leni Riefenstahl, *Leni Riefenstahl: A Memoir* (New York: St. Martin's, 1992), pp. 187–88.

160 "a smart gray flannel": Sylvia Weaver, "Times' Fashion Editor Sees Great Spectacle," *Los Angeles Times*, Aug. 2, 1936, p. A11.

"From noon till night": Thomas Wolfe, *You Can't Go Home Again* (New York: Harper Perennial, 1998), p. 591.

161 "Just twenty-two years ago": Grantland Rice, "Crowd Accords Nations Giving Nazi Salute Thundering Welcome," *Los Angeles Times*, Aug. 2, 1936, p. A11.

"from far away": Frederick T. Birchall, "100,000 Hail Hitler; U.S. Athletes Avoid Nazi Salute to Him," *New York Times*, Aug. 2, 1936, p. 1.

"To Strauss the composer": Norman Stone, "Why the Agony and the Ecstasy?" *Times* (London), June 26, 1988.

162 The music faded: Paul Gallico, "Olympic Fire Re-Lit; Games Begin Today," *New York Daily News*, Aug. 2, 1936, pp. 80, 85.

Meanwhile, Leni Riefenstahl's army: Riefenstahl, *Leni Riefenstahl*, pp. 191–92.

164 The French "had marched": Albert Speer, *Inside the Third Reich* (New York: Touchstone, 1997), p. 73.

166 "This flag dips": The quotation is often erroneously attributed to Martin Sheridan, an enormous Irish-American discus thrower.

167 "It was quite evident": Rice, "Crowd Accords Nations Giving Nazi Salute Thundering Welcome."

"There was a moment": Paul Gallico, "Olympic Fire Borne into Berlin Stadium in Dramatic Rites," *Washington Post*, Aug. 2, 1936, p X1.

"the debate over": Gayle Talbot, "Was It the 'Razz' That American Squad Got?" *Washington Post*, Aug. 2, 1936, p. X1.

168 "I still can't realize": Rice, "Crowd Accords Nations Giving Nazi Salute Thundering Welcome."

170 "demonstration of Nazi organizing": Al Laney, "Nazis' Fervor Finds Climax in Olympic Rites," *New York Herald Tribune*, Aug. 2, 1936, p. 1.

"If he had to have": Larry Snyder, "My Boy Jesse," *Saturday Evening Post*, Nov. 7, 1936, p. 15.

15. DAY ONE

172 "I said, I always": Henry McLemore, "Will Break 3 Records, Says Owens," *Daily Worker*, Aug. 2, 1936, p. 14.

177 "No European crowd": Grantland Rice, "Whirlwind Owens Had Fans Gasping," *Cleveland Plain Dealer*, Aug. 3, 1936.

"I've got to chuck": Cooper C. Graham, *Leni Riefenstahl and Olympia* (Lanham, Md.: Scarecrow, 2001), p. 82.

"have been proclaimed": Grantland Rice, "Berlin at High Pitch," *Los Angeles Times*, Aug. 3, 1936, p. A11.

178 "That doesn't win": Royal Brougham, "Owens Foresees New Mark in Final Today," *New York American*, Aug. 3, 1936.

179 "a funny-looking kid": Dick Schaap, *An Illustrated History of the Olympics* (New York: Knopf, 1975), p. 213.

180 "It isn't for your correspondent": J. P. Abramson, "Owens Beats World Mark in 100-Meter at Olympics," *New York Herald Tribune*, Aug. 3, 1936, p. 1.

"Five minutes before": Arthur J. Daley, "110,000 See Owens Set

World Record at Olympic Games," *New York Times,* Aug. 3, 1936, p. 1.

180   "They just missed": Paul Gallico, "Owens' World-Record Sprint Is Sensation of Olympic Day," *Washington Post,* Aug. 3, 1936, p. X14.

"Since Hitler had": Al Laney, "Hitler Fails to Greet U.S. Negro Winners, but Slur Is Denied," *New York Herald Tribune,* Aug. 3, 1936, p. 16.

181   "Several scientific": Bill Corum, "They Got Rhythm; We Win with 'Em," *New York Journal,* Aug. 3, 1936.

16. DAY TWO

183   "Big, black, awkward yearling": Henry "Stopwatch" McLemore, "Olympic Clocker Tabs Jesse Owens Best Bet," *Los Angeles Times,* Aug. 2, 1936, p. A11.

185   "The American Olympic Committee": Grantland Rice, "Dark Shadow Falls over Herr Hitler as Negro Athletes Dominate Olympics," *Los Angeles Times,* Aug. 4, 1936, p. A13.

186   "As events have turned out": Westbrook Pegler, "Fair Enough," *New York World-Telegram,* Aug. 6, 1936, p. 17.

188   "If I get pressed": Eleanor Holm Jarrett, "Ohio State Lad Pleases Hitler with His Antics," *New York Journal,* Aug. 3, 1936, p. 6.

189   "It wasn't long": Leni Riefenstahl, *Leni Riefenstahl: A Memoir* (New York: St. Martin's, 1992), p. 194.

190   "The stadium was deathly still": Ibid., p. 195.

191   "Metcalfe ran a great race": Edward Beattie, "Happiest Day of His Life for Jesse Owens," *Los Angeles Times,* Aug. 4, 1936, p. A11.

"I'm very glad": *Jesse Owens Returns to Berlin,* directed by Bud Greenspan, 1966 TV documentary.

192   "I can't help wondering": Bill Corum, "They Got Rhythm; We Win with 'Em," *New York Journal,* Aug. 3, 1936.

"Candor compels me": Paul Gallico, "Hitler Waives Jim Crow Law to Extent of Saluting Owens," *Washington Post,* Aug. 4, 1936, p. X15.

Then, after the medalists: William J. Baker, *Jesse Owens: An American Life* (New York: Free Press, 1988), p. 94.

"There was considerable": Gallico, "Hitler Waives Jim Crow Law."

193   "Chancellor Hitler exchanged": Alan Gould, "Owens Wins 100-Meters," *Los Angeles Times,* Aug. 4, 1936, p. A11.

"Hitler's salute to him": William L. Shirer, "Pushed to Limit by Metcalfe, Says Owens," *New York American,* Aug. 3, 1936.

193 "'Mr. Hitler had to'": Louis Effrat, "Owens, Back, Gets Hearty Reception," *New York Times*, Aug. 25, 1936, p. 25. Owens's exact words were reported differently by different reporters. According to Paul Mickelson of the Associated Press, Owens said, "Hitler? Why he was fine. Remember, he was a very busy man. And it seemed he was scheduled to leave the grounds every time I raced. Once, he smiled and waved to me. I waved right back. Stories about him refusing to cheer for American athletes are not true at all."

And the *New York Evening Journal* reported the conversation between Owens and the press yet another way:

> "How about Hitler, Jesse?"
>
> "Well," was the considered reply, "Mr. Hitler had certain definite times to leave the stadium and when I won the 100-meter dash he was just about leaving. He waved at me, though, and I waved at him . . . I think it would be bad taste to criticize Germany's man of the hour."

"Hitler Waives": Gallico, "Hitler Waives Jim Crow Law."

"Jesse swears": Larry Snyder, "My Boy Jesse," *Saturday Evening Post*, Nov. 7, 1936, p. 97.

194 "Herr Hitler, Nazi dictator": "Hitler Snubs Jesse," *Cleveland Call and Post*, Aug. 6, 1936, p. 1.

"It has been demonstrated": "No Snub Was Intended," *Pittsburgh Courier*, Aug. 8, 1936, p. 1.

"has captured everyone": "World's Sports King," *Chicago Defender*, Aug. 8, 1936, p. 1.

"There has been a cloud": "Hitler . . . and Jesse Owens," *New York Daily News*, Aug. 6, 1936, p. 27.

195 "Each time Owens trotted up": William L. Shirer, *The Nightmare Years, 1930–1940* (Edinburgh: Birlinn, 2001), p. 234.

"a handsome young man": Ibid.

"The Americans ought to be ashamed": Ibid.

17. DAY THREE

197 "I think maybe": "Jesse Stars Today in 200 Meters and Broad Jump," *New York Journal*, Aug. 4, 1936.

198 "most certain": Arthur J. Daley, "U.S. Captures 4 Events," *New York Times*, Aug. 5, 1936, p. 1.

And Alan Gould: Alan Gould, "Owens Sets 2 Records," *Washington Post*, Aug. 5, 1936, p. X1.

199 "Long was one": Jesse Owens, *Jesse: The Man Who Outran Hitler* (New York: Fawcett, 1985), p. 62.

201 The American custom: Years later, for reasons unknown, Owens wrote that on his first attempt he simply took off beyond the board and fouled. But several thousand eyewitnesses saw him simply run through the pit in his sweat suit.

202 "The situation": Daley, "U.S. Captures 4 Events."

203 "What has taken your goat": Owens, *Jesse*, p. 72.

"You know, you should": Dick Schaap, *An Illustrated History of the Olympics* (New York: Knopf, 1975), p. 211.

204 "some telltale sign of emotion": William J. Baker, *Jesse Owens: An American Life* (New York: Free Press, 1988), p. 97.

"See": Long's gesture went unnoticed by the press corps. Summing up Owens's qualifying troubles, Paul Gallico wrote, "Owens scared everyone, including himself, by running through his first trial as a warm-up, only to find it counted as one. The second time he stepped over the take-off and that left but one more trial. He took no chances on this one and took off a foot behind the white mark and did 25 feet" ("U.S. Wins 4 Olympic Titles," *New York Daily News*, Aug. 5, 1936, p. 52). Gallico made no mention of Luz Long or his subsequently famous gesture. Neither did Daley or Gould.

205 "they had foreseen": "Olympic Games," *Time*, Aug. 17, 1936.

Gender uncertainty was actually: "Olympic Games," *Time*, Aug. 24, 1936.

206 "reviving charges Owens": Davis J. Walsh, *New York Journal*, Aug. 4, 1936, p. 22.

"Boy, what a thrill": International News Service, "Hitler Handshake after Victory Leaves Stephens Trembling," *New York Journal*, Aug. 4, 1936.

207 "It is something": Grantland Rice, "Jesse Owens, Woodruff Steal Show," *Los Angeles Times*, Aug. 5, 1936, p. A15.

At 4:30, in the round: Baker, *Jesse Owens*, p. 97.

208 "His eagerness to receive": Daley, "U.S. Captures 4 Events."

"If America didn't have": "Innuendo by Nazis Arouses Catholics," *New York Times*, Dec. 17, 1936, p. 14.

209 "Tuesday was a dark": Rice, "Jesse Owens, Woodruff Steal Show."

210 "It begins to look": Joe Williams: "Negro Stars Shine in Games, Give America Lead in Points, No More Hitler Greetings," *New York World-Telegram*, Aug. 4, 1936, p. 22.

"I haven't even thought": Grantland Rice, *The Tumult and the Shouting* (New York: A. S. Barnes, 1963), p. 253.

210 "The crowning achievements": "Nazi Insults to Negro Stars Condemned by Herndon," *Daily Worker*, Aug. 12, 1936, p. 1.

211 "Hitler didn't snub me": United Press, "Snubbed by Roosevelt, Not Hitler, Says Owens," Oct. 16, 1936.

"The games were overshadowed": Thomas Wolfe, *You Can't Go Home Again* (New York: Harper Perennial, 1998), p. 589.

"Everywhere the air was filled": Ibid., p. 590.

18. "HE FLIES LIKE THE *HINDENBURG*": DAY FOUR

214 "It is my pleasure": Associated Press, "Gov. Daley of Ohio Cables Felicitations to Owens," *New York Times*, Aug. 5, 1936, p. 27.

"Hitler declared Aryan supremacy": Shirley Povich, "This Morning . . . ," *Washington Post*, Aug. 5, 1936, p. X18.

"They are great": "Owens Runs Like the Hindenburg Flies — Schmeling," *Pittsburgh Courier*, Aug. 15, 1936.

215 "America's 'athletes of bronze'": Robert L. Vann, "Hitler Salutes Jesse Owens," *Pittsburgh Courier*, Aug. 8, 1936, p. 1.

"Though it can't be": Genet, "Berlin Letter," *The New Yorker*, Aug. 15, 1936.

"I am proud": Robert L. Vann, "Proud I'm an American, Owens Says," *Pittsburgh Courier*, Aug. 8, 1936, p. 1.

216 To be sure, Mack Robinson: Frank Litsky, "Mack Robinson, 85, Second to Owens in Berlin," *New York Times*, Mar. 14, 2000, p. C30.

217 "[Owens] looked like a dark streak": Grantland Rice, "Foreign Athletes Goggle-Eyed as Owens Achieves His Olympic Games 'Triple,'" *Los Angeles Times*, Aug. 6, 1936, p. A11.

Tens of thousands of Germans: William J. Baker, *Jesse Owens: An American Life* (New York: Free Press, 1988), p. 101.

"If the Olympics clearly": Frederick T. Birchall, "First Yacht Event Delayed by Storm; But Wind Moderates and Race Starts With Olympic Fire Burning on One Ship," wireless to *New York Times*, Aug. 5, 1936, p. 25.

218 "Owens was black as tar": Baker, *Jesse Owens*, p. 101.

"Incomparable," he wrote: Alan Gould, "Owens, Carpenter, Meadows Break Records," *Los Angeles Times*, Aug. 6, 1936, p. A9.

"I told you": Rice, "Foreign Athletes Goggle-Eyed."

"I'm just getting the feel": Associated Press, "Owens Would Race with U.S. Relay," *Washington Post*, Aug. 6, 1936, p. X17.

19. THE RELAY

219 "Jesse Owens's Olympics": Joe Williams, "Jesse Owens Finishes Task; Other Nations Get Chance; British Receive Bad Shock," *New York World-Telegram*, Aug. 6, 1936, p. 20.

"Owens has had enough glory": Associated Press, "Owens Would Race with U.S. Relay," *Washington Post*, Aug. 6, 1936, p. X17.

220 "Jesse, who hates to stand around": Associated Press, "Owens Out of Relay," *New York Times*, Aug. 5, 1936, p. 27,

"Taking the 200-meter run": John Kieran, "Sports of the Times: Three for the Streak," *New York Times*, Aug. 6, 1936, p. 28.

"decided definitely tonight": Associated Press, "Owens Named on 400-Meter Relay Team as Reports of German Strength Alarm Coaches," *New York Herald Tribune*, Aug. 8, 1936, p. 15.

221 "their expected assignments": Ibid.

222 By that time, everyone assumed: Associated Press, "Glickman Says Picking of Team 'Cromwell's Influence,'" *Washington Post*, Aug. 9, 1936, p. X1.

223 "Boys, this is a tough decision": William O. Johnson, *All That Glitters Is Not Gold: The Olympic Games* (New York: Putnam, 1972), p. 179.

"I've got my medals": Ibid., p. 180.

224 "Jesse is one of my best friends": Ibid., p. 181.

"In justice": J. P. Abramson, "Americans Equal Figures in 100-Meter Relay Test," *New York Herald Tribune*, Aug. 9, 1936.

"Originally, it was definitely": Ibid.

225 "The heralded Dutch and German": Associated Press, "Morris Breaks Record as U.S. Sweeps Decathlon," *Chicago Daily Tribune*, Aug. 9, 1936, p. A1.

"the nervous tension": Associated Press, "Owens 'Shaky' as Mobs Clamor for Autographs," *New York Times*, Aug. 9, 1936, p. S2.

226 "The heats failed to show": Associated Press, "Glickman Says Picking of Team 'Cromwell's Influence.'" Even though Glickman and Stoller were denied their gold-medal opportunity, the Germans were forced to present Jewish athletes with fourteen medals, including nine that were gold. Thirteen different Jews won medals in Berlin; one, Endre Kabos of Hungary, a saber specialist, won two gold medals.

227 "I held out my hand": Leni Riefenstahl, *Leni Riefenstahl: A Memoir* (New York: St. Martin's, 1992), p. 196.

"strangely convoluted strokes": Ibid., pp. 200–201.

228 "Owens lit out": Paul Gallico, "Jap Captures Marathon," *New York Daily News,* Aug. 10, 1936, p. 38.

"that the white boys": Paul Gallico, *A Farewell to Sport* (New York: Knopf, 1938), p. 309.

229 "The 400-meter relay was the romp": J. P. Abramson, "250,000 Watch Kitei Son, Corean, Capture Olympic Marathon for Japan in Record Time," *New York Herald Tribune,* Aug. 10, 1936, p. 16.

EPILOGUE

230 "Did you see the games": Joseph C. Nichols, "Schmeling Visits Joe Louis's Camp," *New York Times,* Aug. 10, 1936, p. 15.

231 "I'm anxious to finish": Associated Press, "Owens to Turn Pro If Offers Suit Him," *New York Times,* Aug. 11, 1936, p. 26.

Eddie Cantor: "Czar Brundage Suspends Owens," *New York Daily News,* Aug. 17, 1936, p. 38.

"It would be foolish": Associated Press, "Metcalfe Beats Jesse Owens in Exhibition Race," *Chicago Daily Tribune,* Aug. 11, 1936, p. 22.

232 "I don't know": Associated Press, "Owens Dazzled by Many Offers," *New York Times,* Aug. 16, 1936, p. S2.

"We had no alternative": "Owens Is Barred by AAU; Ready to Become a Pro," *Chicago Daily Tribune,* Aug. 17, 1936, p. 19.

"Last Sunday," he said: Associated Press, "Owens to Wait Until He Returns Home Before Making Decision on Pro Offers," *New York Times,* Aug. 18, 1936, p. 23.

233 "There's nothing I can": Associated Press, "'Unfair,' Says Snyder," *Chicago Daily Tribune,* Aug. 17, 1936, p. 19.

"It is pretty terrible": Associated Press, "'Scandalous,' Says Jesse's Ma," *Los Angeles Times,* Aug. 18, 1936, p. A10.

For three days: Donald McRae, *Heroes Without a Country: America's Betrayal of Joe Louis and Jesse Owens* (New York: Ecco, 2002), p. 169.

235 "My heart is telling me": E-mail from Jorg Weck, archivist, German Sport and Olympic Museum, Cologne, May 11, 2005.

"I've seen Luz again": Ibid.

Acknowledgments

Triumph could not have been written without benefit of the talents of the men who in the mid-1930s covered Jesse Owens, the Olympic boycott movement, and the games of the Eleventh Olympiad: Grantland Rice, Paul Gallico, Westbrook Pegler, Alan Gould, Braven Dyer, Arthur J. Daley, John Kieran, Bill Corum, Frederick T. Birchall, Al Laney, Jimmy Cannon, Joe Williams, Lewis Burton, Shirley Povich, Henry McLemore, R. Walter Merguson, Jesse Abramson, Fred Farrell, and William L. Shirer. Their colorful prose and courageous reporting painted a vivid picture of Owens and his time. While they have all died — as have many of the newspapers for which they wrote — some are well remembered, even legendary, but others are all but forgotten. They were all invaluable to me.

So were several anonymous reporters and columnists at the major black newspapers, whose names I do not know because their stories rarely carried bylines. To these writers at the *Atlanta Daily World,* the *Cleveland Call and Post,* the *Pittsburgh Courier,* the *Chicago Defender,* and the *Amsterdam News,* I offer my thanks.

I am also deeply indebted to several other writers and historians. There is no better account of the life of Jesse Owens than Professor William J. Baker's *Jesse Owens: An American Life,* from which I culled a tremendous amount of information. Owens's three autobiographies, written with the assistance of Paul Neimark, were also fonts of anecdotes and recollections, as was William Oscar Johnson's *All That Glitters Is Not Gold.* I repeatedly turned to several other sources: Donald McRae's *Heroes Without a Country;* Richard Mandell's *The Nazi Olympics;* Duff Hart-

Davis's *Hitler's Games: The 1936 Olympics;* Cooper C. Graham's *Leni Riefenstahl and Olympia;* Frank Deford's profile of Leni Riefenstahl in *Sports Illustrated;* Larry Snyder's article "My Boy Jesse" in the *Saturday Evening Post;* Leni Riefenstahl's *A Memoir* and her film *Olympia;* David Wallechinsky's *The Complete Book of the Summer Olympics;* Bud Greenspan's documentary *Jesse Owens Returns to Berlin;* Grantland Rice's *The Tumult and the Shouting;* Paul Gallico's *A Farewell to Sport;* Susan D. Bachrach's *The Nazi Olympics: Berlin 1936;* William L. Shirer's *The Nightmare Years, The Rise and Fall of the Third Reich,* and *Berlin Diary;* Thomas Wolfe's *You Can't Go Home Again;* Frederick Spotts's *Hitler and the Power of Aesthetics;* Marty Glickman and Stan Isaacs's *The Fastest Kid on the Block;* Victor Klemperer's *I Will Bear Witness;* Albert Speer's *Inside the Third Reich;* Robert Caro's *The Power Broker;* Dick Schaap's *An Illustrated History of the Olympics;* and *The Encyclopedia of New York City,* edited by Kenneth T. Jackson.

Also of enormous value to me were the archives of the New York Public Library's Schomburg Center for Research in Black Culture, the Western Reserve Historical Society, the Cleveland Public Library, the *New York Times, Time* magazine, the Hearst syndicate, the *Chicago Tribune,* the *Washington Post,* the *Los Angeles Times,* the *New York Herald Tribune,* the *New York Daily News,* and the *Cleveland Plain Dealer,* as was the collection of the Museum of Television and Radio.

I have patched together many of the stories in the book from not one but several of the aforementioned sources. For instance, there are dozens of differing accounts of the events of August 4, 1936, when Luz Long made his unforgettable gesture of sportsmanship during the Olympic broad-jump competition. Arthur Daley's version is slightly different from Paul Gallico's, which is slightly different from Jesse Abramson's. Additionally, Jesse Owens's personal recollections are often at odds not only with the historical record but with each other. His account of the so-called Hitler snub, for example, changed frequently over the decades. In the interest of historical accuracy and narrative flow, I have relied on the versions of events offered by him and others that I found most credible.

Regarding dialogue, I have tried to use wherever possible the exact words recorded by the reporters who were writing them down as they were spoken. But on occasion I have, necessarily, taken some license in embellishing and editing the dialogue. I have endeavored to remain true to the essential facts and convictions of the men and women who grace these pages, and always I have done so only in service to the story.

This book owes its very existence to the brilliance of Susan Canavan, my editor at Houghton Mifflin. Her vision, dedication, and talent are unmatched. I stand in awe of her ability to fashion a book from my rambling prose — though I still think that Will Rogers deserved a few more paragraphs.

Scott Waxman, my literary agent, first attracted me to the idea of writing a book about Jesse Owens. He is the best at what he does, a tireless advocate and creative dynamo.

I am also grateful for the contributions of Martha Kennedy, who designed the stirring cover; Liz Duvall, the ever-exacting high priestess of red pencil, who cleaned up my sentences; Melissa Lotfy, who set those clean sentences into beautifully designed type; Laura Noorda, keeper of the schedule, who somehow managed to pull together this project on time despite my endless tinkering; Deborah DeLosa, Houghton Mifflin's well-connected publicist; Sanj Kharbanda, the house marketing guru; Greg Payan, my diligent and keen-eyed photo researcher; and Sarah Gabert, the master of the notes.

I must also thank my dedicated researcher, the incomparable Joe Goldstein. He gathered source material everywhere from Cologne to Los Angeles, cashing in thousands of frequent-flier miles and innumerable markers with his contacts in the world of newspapers and magazines. Meanwhile, Willie Weinbaum, my esteemed colleague at ESPN, lent his considerable talents to both the fact-gathering and the fact-checking process. His unerring eye for detail saved me from myself more than once. Dave Smith of the New York Public Library delved deep into that formidable institution's recesses to locate obscure newspaper and magazine stories that enriched this book. He is an institution himself. Few people are better versed on the fine points of track and field in the 1930s than Marvin Rothenstein, whose expertise I abused.

Similarly, Bill Mallon—surgeon, golfer, and Olympic historian —allowed me to tap his remarkable brain. Rebecca Aronauer and Marc Aronin, my trusty interns, gave me some of their finest hours, with no college credits to be gained. Tamar Chute of the Ohio State University archives provided valuable support, affording me access to the archives' vast catalogue of Owens documentation. Ray Lumpp of the New York Athletic Club, Randy Harvey of the *Los Angeles Times*, Ryan Schiavo of ESPN, Professor Pam Laucella of Indiana University, Wayne Wilson of the Amateur Athletic Foundation of Los Angeles, and Jorg Weck of the German Sport and Olympic Museum in Cologne generously shared their resources and time. Olympic champions John Woodruff and Harrison Dillard and Olympic bronze medalist Herb Douglas graciously submitted to interviews and spoke on the topic of their friend Jesse Owens. Kai Long spoke about the father he never got to know. Ralph Metcalfe, Jr., and Dave Wykoff were kind enough to discuss their fathers. Ira Berkow and Frank Litsky, two distinguished gentlemen of the *New York Times*, lent their memories and insights as well.

A special thanks is owed Gina Hemphill, Jesse Owens's granddaughter, and the rest of the Owens family. Their support and generosity were crucial to this endeavor. My interactions with them over the years piqued my interest in the exploits of the family patriarch.

Thanks also to Will Vincent, Mike Lupica, Jimmy Breslin, Sandy Padwe, Farley Chase, Ellen Harilal, Dave Herscher, Vince Doria, Gare Joyce, Mark Moore, Sarah Starr, and Dave Zirin.

Finally, I would like to acknowledge the influence of my father, Dick Schaap. He died in 2001, long before this book was conceived, but it never would have been written had he not nurtured my enthusiasm for all things Olympic. We were together at four Olympics, in Albertville, Barcelona, Lillehammer, and Atlanta. More would have been nice.

Index